X-RAY TECHNOLOGY EXAMINATION REVIEW BOOK

Volume 2

Third Edition

P9-BZR-997

Edited by

WILLIAM O. CRAWFORD, JR., M.D.

Assistant Professor of Radiology

Department of Radiology

Yale University School of Medicine

Yale University

New Haven, Connecticut

HENRI J. GAUTOT, R.T., A.R.R.T.

Director of Technical Education

Department of Radiology

Yale - New Haven Hospital

New Haven, Connecticut

1500 MULTIPLE CHOICE QUESTIONS AND ANSWERS REFERENCED TO TEXTBOOKS

MEDICAL EXAMINATION PUBLISHING COMPANY, INC.

65-36 Fresh Meadow Lane

Flushing, N.Y. 11365

Library of Congress
Catalog Card Number

64-17385

ISBN 0-87488-442-X

October, 1973

PRINTED IN THE UNITED STATES OF AMERICA

PREFACE

The purpose of this text is to encourage the reader to detect areas of weakness in his understanding of subject matter so that he may refer to his texts for a more comprehensive review of the subject. It also will provide an interesting challenge to the student, as well as an opportunity to improve his skills with multiple-choice examinations.

The completely referenced form of the question material is provided to facilitate the reader in checking his areas of weakness. This book will prove to be a time-saving device in rapidly finding a source of information. While testing your memory be aware of the textbook as being the ultimate source of your fund of knowledge. References cited in the individual questions are listed in the back of the book.

The variations in test questions are offered to familiarize the reader with various types of National Board, State Board and School examinations in use today. The following test material will give you an opportunity to determine your ability to read, digest and comprehend the vast accumulation of available knowledge. Do not be discouraged if you are unable to achieve a perfect test score; indeed be stimulated to learn the material by referring to your textbooks. Successfully solving the problems in this text will surely give you satisfaction.

X-RAY TECHNOLOGY
EXAMINATION REVIEW BOOK
VOLUME 2
Third Edition

TABLE OF CONTENTS

FOR EACH OF THE FOLLOWING MULTIPLE CHOICE QUESTIONS
SELECT THE ONE MOST APPROPRIATE ANSWER:

1. Enlargement of lymph nodes is referred to as:
    A. Abscess
    B. Adenopathy
    C. Adenoma
    D. Adenoid   Ref. 1 - p. 76

2. Fibrous bands between two normally separated structures are known as:
    A. Adhesions
    B. Atelectasis
    C. Cellulitis
    D. Vesicles   Ref. 1 - p. 76

3. Deficiency in quantity or quality of blood is:
    A. Anemia
    B. Hemorrhage
    C. Sarcoma
    D. Hemophilia   Ref. 1 - p. 100

4. An abnormal swelling or dilatation of artery due to weakness of its walls
    is called:
    A. Insufficiency
    B. Atherosclerosis
    C. Aneurysm
    D. Coarctation   Ref. 1 - p. 103

5. Collapse or loss of air in lung tissue is known as:
    A. Pneumothorax
    B. Emphysema
    C. Effusion
    D. Atelectasis   Ref. 1 - p. 76

6. A condition of blood vessels often manifested by calcification of their
    walls is referred to as:
    A. Hypercalcemia
    B. Atherosclerosis
    C. Hypertension
    D. Aortic stenosis   Ref. 1 - p. 104

7. Abnormal union and immobility of the bones of a joint is called:
    A. Sprain
    B. Arthritis
    C. Luxation
    D. Ankylosis   Ref. 1 - p. 44

8. Enlargement of the hands, feet, jaw and thorax due to increased
    secretion of growth hormone is known as:
    A. Achondroplasia
    B. Acrocyanosis
    C. Acromegaly
    D. Arachnodactyly   Ref. 1 - p. 48

9. Disease of joints is usually called:
    A. Arthritis
    B. Atheroma
    C. Osteomyelitis
    D. Osteitis   Ref. 1 - p. 48

10. A skull that is short from front to back and wide from side to side is:
    A. Bicornuate
    B. Brachycephalic
    C. Dolicocephalic
    D. Biparietal   Ref. 1 - p. 31

11. An unexpected, exaggerated, life threatening reaction to the injection of drugs or contrast agents is known as:
    A. Anaphylaxis
    B. Analepsis
    C. Allergy
    D. Anergy   Ref. 2 - p. 78

12. Division into two parts is called:
    A. Canulation
    B. Canalyzation
    C. Bifurcation
    D. Bilateral   Ref. 1 - p. 207

13. Persistent dilatation of bronchi caused by infection is known as:
    A. Bronchopneumonia
    B. Pneumonitis
    C. Bronchiolitis
    D. Bronchiectasis   Ref. 1 - p. 82

14. Callus is a substance formed around bone in response to:
    A. Fracture
    B. Againg
    C. Tumor
    D. Inflammation   Ref. 1 - p. 44

15. Cardiomegaly means:
    A. Myocardial infarction
    B. Congestive heart failure
    C. Coronary artery disease
    D. Cardiac enlargement   Ref. 1 - p. 101

16. Merging structures or disease processes are called:
    A. Convergent
    B. Confluent
    C. Consolidated
    D. Discrete   Ref. 1 - p. 79

17. Cataract is an opacity of the _____ of the eye:
    A. Lens
    B. Cornea
    C. Conjunctiva
    D. Retina   Ref. 1 - p. 226

18. The prefix angio - denotes relationship to:
    A. Urinary tract
    B. Blood and lymph vessels
    C. Gastrointestinal tract
    D. Central nervous system   Ref. 2 - p. 88

19. Hyaline cartilage calcification is a manifestation of:
    A. Chondrocalcinosis
    B. Scleroderma
    C. Calcinosis circumscripta
    D. Arteriosclerosis   Ref. 1 - p. 49

20.   Skull deformity secondary to early fusion cranial sutures is called:
       A.   Craniotomy
       B.   Hydrocephalus
       C.   Craniomegaly
       D.   Craniostenosis   Ref. 1 - p. 49

21.   Cirrhosis usually refers to disease of the:
       A.   Pancreas
       B.   Heart
       C.   Liver
       D.   Spleen   Ref. 1 - p. 131

22.   Solidification of lung tissue due to displacement of alveolar air by edema
       fluid or pus is called:
       A.   Caseation
       B.   Consolidation
       C.   Cavitation
       D.   Effusion   Ref. 1 - p. 79

23.   Location of the heart on the right side instead of the left side of the chest
       is known as:
       A.   Dextrocardia
       B.   Cor Bovinum
       C.   Levocardia
       D.   Cardiomyopathy   Ref. 1 - p. 101

24.   Separation of two structures normally joined together is called:
       A.   Dystrophy
       B.   Diastasis
       C.   Dysplasia
       D.   Hypoplasia   Ref. 1 - p. 45

25.   Enteritis refers to inflammation of the:
       A.   Heart
       B.   Pharynx
       C.   Arteries
       D.   Intestine   Ref. 1 - p. 125

26.   Ivory like change in bone is known as:
       A.   Eventratum
       B.   Extravasation
       C.   Eburnation
       D.   Ectopia   Ref. 1 - p. 45

27.   Dilatation of peripheral air spaces in the lungs is called:
       A.   Endocarditis
       B.   Cor pulmonale
       C.   Emphysema
       D.   Bronchiolitis   Ref. 1 - p. 84

28.   Empyema is accumulation of pus in the:
       A.   Pleural space
       B.   Bronchi
       C.   Pericardial space
       D.   Peritoneal cavity   Ref. 1 - p. 79

29. A marked deviation from the normal, such as six fingers on one hand, is known as a(an):
   A. Anopsia
   B. Neoplasm
   C. Atrophy
   D. Anomaly   Ref. 2 - p. 98

30. Effusion of fluid into the abdominal cavity is called:
   A. Ascites
   B. Peritonitis
   C. Pyuria
   D. Empyema   Ref. 2 - p. 151

31. Loss of feeling or sensation is called:
   A. Hypoxia
   B. Analepsis
   C. Anesthesia
   D. Asphyxia   Ref. 2 - p. 85

32. Hormones are secreted by _____ glands:
   A. Sebaceous
   B. Exocrine
   C. Lymph
   D. Endocrine   Ref. 1 - p. 172

33. Esophageal varices are enlarged tortuous _____ in the esophagus:
   A. Arteries
   B. Veins
   C. Glands
   D. Lymphatics   Ref. 1 - p. 128

34. A broncholith is a _____ within the bronchial tree:
   A. Calculus
   B. Gland
   C. Infection
   D. Tumor   Ref. 2 - p. 222

35. The term cachexia means:
   A. Good health and nutrition
   B. Profound poor health and malnutrition
   C. Obesity
   D. Shortness of breath   Ref. 2 - p. 233

36. Goiter is a term meaning enlargement of the:
   A. Thyroid
   B. Thymus
   C. Pancreas
   D. Heart   Ref. 1 - p. 179

37. The term carpopedal pertains to:
   A. Hand and shoulder
   B. Wrist and foot
   C. Wrist and heel
   D. Wrist and ankle   Ref. 2 - p. 256

38. Excess amount of cerebrospinal fluid in ventricles of the brain is called:
   A. Hypophysis
   B. Hydramnios
   C. Hydrocele
   D. Hydrocephalus   Ref. 1 - p. 200

39. Dilatation of the esophagus due to persistent contraction of the gastro-esophageal junction is called:
A. Dyspepsia
B. Presbyesophagus
C. Achalasia
D. Chalasia   Ref. 1 - p. 124

40. When speaking of a clover leaf deformity, reference is made to deformity due.to:
A. Colitis
B. Gastric ulcer
C. Duodenal ulcer
D. Diverticulitis   Ref. 1 - p. 124

41. Acute angulation of the spine due to collapse of a vertebral segment is called:
A. Gibbus
B. Lordosis
C. Spondylosis
D. Scoliosis   Ref. 1 - p. 46

42. A glioma is a tumor of the:
A. Kidney
B. Brain
C. Pancreas
D. Skin   Ref. 1 - p. 203

43. Obstruction of the urinary tract causing dilation of renal pelvis and calyces is called:
A. Hydrocele
B. Pyonephrosis
C. Hydronephrosis
D. Hydrope   Ref. 1 - p. 228

44. Hydrosalpinx refers to distention of _____ with fluid:
A. Colon
B. Gall bladder
C. Urinary bladder
D. Fallopian tube   Ref. 1 - p. 158

45. An area of tissue necrosis resulting from circulatory insufficiency is called an:
A. Infarct
B. Occlusion
C. Involution
D. All of the above   Ref. 1 - p. 102

46. Short transverse lines appearing in the periphery of the lung bases in congestive heart failure are known as:
A. Failure lines
B. Kerley B lines
C. Kerley C lines
D. Twining's lines   Ref. 1 - p. 81

47. The term linitis plastica refers to tumor infiltration of the wall of the _____ making it rigid:
A. Stomach
B. Duodenum
C. Ileum
D. Jejunum   Ref. 1 - p. 139

48.   Pseudofractures of bone are called:
    A.   Twining's lines
    B.   Alveolar lines
    C.   Looser's lines
    D.   Poupart's lines   Ref. 1 - p. 46

49.   A mesothelioma is a tumor of:
    A.   Meninges
    B.   Pleura
    C.   Pericardium
    D.   Synovia   Ref. 1 - p. 90

50.   Narrowing of the valve between the left atrium and left ventricle is
    called mitral _____:
    A.   Stenosis
    B.   Insufficiency
    C.   Regurgitation
    D.   Prolapse   Ref. 1 - p. 108

51.   Radiographically dense but structurally fragile bones are found in:
    A.   Osteomyelitis
    B.   Osteomalacia
    C.   Osteopetrosis
    D.   Osteoporosis   Ref. 1 - p. 59

52.   Papillary muscles are found in the:
    A.   Bronchi
    B.   Tongue
    C.   Fingers
    D.   Heart   Ref. 1 - p. 98

53.   Pheochromocytoma is a tumor of the:
    A.   Pancreas
    B.   Adrenal glands
    C.   Thyroid gland
    D.   Pituitary gland   Ref. 1 - p. 183

54.   Flattening of the base of the skull is called:
    A.   Brachycephaly
    B.   Platybasia
    C.   Platyspondyly
    D.   Dolicocephaly   Ref. 1 - p. 47

55.   Inflammation of the gall bladder is called:
    A.   Cholelithiasis
    B.   Cholecystitis
    C.   Pyelitis
    D.   Cystitis   Ref. 2 - p. 295

56.   A fever producing substance is known as a(n):
    A.   Chromogen
    B.   Microorganism
    C.   Pyrogen
    D.   Antigen   Ref. 1 - p. 230

57.   The term rachitic rosary refers to abnormality of:
    A.   Phalanges
    B.   Ribs
    C.   Vertebrae
    D.   Metacarpals   Ref. 1 - p. 47

58. Silicosis is a disease of:
     A. Stomach
     B. Skin
     C. Heart
     D. Lungs   Ref. 1 - p. 86

59. Inflammation of vertebrae is called:
     A. Spondylosis
     B. Spondylitis
     C. Arachnoiditis
     D. Spondylolysis   Ref. 1 - p. 48

60. Cystitis refers to inflammation of the:
     A. Urinary bladder
     B. Gall bladder
     C. Pyelitis
     D. Cystitis   Ref. 2 - p. 378

61. Incomplete dislocation of joints is called:
     A. Subluxation
     B. Spondylolisthesis
     C. Syndactyly
     D. Spondylosis   Ref. 1 - p. 48

62. Sclerosis refers to:
     A. Softening
     B. Melting
     C. Freezing
     D. Hardening   Ref. 1 - p. 48

63. Scurvy is a disease resulting from deficiency of:
     A. Iron
     B. Vitamin B_{12}
     C. Vitamin D
     D. Vitamin C   Ref. 1 - p. 64

64. Syringomyelea is a condition marked by progressive enlargement of
     fluid filled cysts within the:
     A. Muscles
     B. Sciatic nerve
     C. Spinal cord
     D. Brain   Ref. 1 - p. 195

65. Surgical puncture of the chest wall with a needle for the purpose of
     withdrawing fluid from the pleural space is called:
     A. Thoracentesis
     B. Thoracotomy
     C. Thoracoplasty
     D. Thoracostomy   Ref. 1 - p. 82

66. The tearing of a ligament is called a:
     A. Stress fracture
     B. Torus fracture
     C. Sprain
     D. Strain   Ref. 1 - p. 54

67. Truncus arteriosus is a congenital abnormality of the:
     A. Thorax
     B. Stomach
     C. Heart
     D. Kidneys   Ref. 1 - p. 106

68. Enlarged tortuous veins are called:
    A. Varicosities
    B. Aneurysms
    C. Variants
    D. Ventricles   Ref. 1 - p. 103

69. Wilm's tumor is a neoplasm of the _____, affecting primarily children under the age of seven:
    A. Colon
    B. Kidney
    C. Bladder
    D. Brain   Ref. 1 - p. 170

70. A fragment of dead bone in an area of bone destruction is called a(n):
    A. Rarefaction
    B. Sequestrum
    C. Involucrum
    D. Nidus   Ref. 1 - p. 48

71. A blood vessel deviating from its normal course is called:
    A. Aberrant
    B. Varicose
    C. Abortive
    D. Devious   Ref. 2 - p. 2

72. Absence of acid secretions in the stomach is known as:
    A. Achromasia
    B. Chlorosis
    C. Achalasia
    D. Achlorhydria   Ref. 2 - p. 13

73. Spasmodic hysterical air swallowing is called:
    A. Aerophagia
    B. Deglutition
    C. Flatulence
    D. None of the above   Ref. 2 - p. 42

74. The surgical formation of a passage between two normally separate organs is known as a(n):
    A. Adhesion
    B. Ankylosis
    C. Junction
    D. Anastomosis   Ref. 2 - p. 79

75. The term buccal pertains to the:
    A. Lips
    B. Cheek
    C. Nose
    D. Chin   Ref. 2 - p. 225

76. Inflammation of the skin is called:
    A. Conjunctivitis
    B. Cystitis
    C. Proctitis
    D. Dermatitis   Ref. 2 - p. 401

77. A fracture resulting in many bone fragments is called:
    A. Distracted
    B. Impacted
    C. Comminuted
    D. Compound   Ref. 2 - p. 330

78. The study of cells is called:
    A. Mycology
    B. Cytology
    C. Virology
    D. All of the above   Ref. 2 - p. 382

79. Endothelium is a layer of cells lining:
    A. Cerebral ventricles
    B. Stomach
    C. Blood vessels
    D. Bronchi   Ref. 2 - p. 491

80. The anatomic term denoting a hollow or depressed area is:
    A. Foramen
    B. Fornix
    C. Fistula
    D. Fossa   Ref. 2 - p. 578

81. Accumulation of blood in a joint is called:
    A. Hematemesis
    B. Hemarthrosis
    C. Hematobilia
    D. Hematocrit   Ref. 2 - p. 653

82. Dilatation or distention of a structure is called:
    A. Aphasia
    B. Ectasia
    C. Acalasia
    D. Dysphagia   Ref. 2 - p. 465

83. Removal and examination of a piece of body tissue for diagnostic
    purposes is:
    A. Biopsy
    B. Histology
    C. Pathology
    D. Incision   Ref. 2 - p. 198

84. Congenital absence of a body structure or organ is called:
    A. Atrophy
    B. Achalasia
    C. Agenesis
    D. Hypertrophy   Ref. 2 - p. 46

85. The presence of abnormally large amounts of fluid in body tissues is
    known as:
    A. Edema
    B. Effusion
    C. Infusion
    D. Inflammation   Ref. 2 - p. 467

86. The type of aneurysm in which blood is forced between the coats of an
    artery is called a _____ aneurysm:
    A. Berry
    B. Cirsoid
    C. Diffuse
    D. Dissecting   Ref. 2 - p. 87

87.   An absence or lack of oxygen is referred to as:
        A.   Aphasia
        B.   Anoxia
        C.   Oxydesis
        D.   Anaphylaxis   Ref. 2 - p. 100

88.   A substance which destroys bacteria and is therefore used to treat
        infection is called a(an):
        A.   Antibiotic
        B.   Antispasmodic
        C.   Anticoagulant
        D.   Antidiabetic   Ref. 2 - p. 104

89.   Glaucoma is a disease of which part of the body?:
        A.   Heart
        B.   Lung
        C.   Eye
        D.   Ear   Ref. 2 - p. 617

90.   A _____ is an abnormal passage between two organs or between an
        internal organ and the skin:
        A.   Fissure
        B.   Sinus
        C.   Foramen
        D.   Fistula   Ref. 2 - p. 561

91.   Cystectomy refers to surgical removal of the:
        A.   Lung
        B.   Bladder
        C.   Intestine
        D.   Kidney   Ref. 2 - p. 378

92.   A bluish discoloration of the skin and mucous membranes due to a lack of
        oxygen and/or hemoglobin in the blood is called:
        A.   Cyanosis
        B.   Cyanide
        C.   Hemophilia
        D.   Oxidation   Ref. 2 - p. 372

93.   A condition caused by obstruction of the larynx, occurring mainly in
        children and characterized by a barking cough, hoarseness and difficulty
        in breathing is called:
        A.   Pneumonia
        B.   Measles
        C.   Croup
        D.   Atelectasis   Ref. 2 - p. 359

94.   A craniotomy refers to an operation on the:
        A.   Spine
        B.   Skull
        C.   Gall bladder
        D.   Mandible   Ref. 2 - p. 354

95.   Deficiency of blood in a portion of the body is known as:
        A.   Ischemia
        B.   Hyperemia
        C.   Ischium
        D.   Hemoglobinuria   Ref. 2 - p. 760

96. Barium that is inspissated in the colon is:
    A. Thin and watery
    B. Thick and dried
    C. A desired condition
    D. Powdery   Ref. 2 - p. 745

97. An indurated area is:
    A. Soft
    B. Liquid
    C. Hardened
    D. Dried out   Ref. 2 - p. 737

98. The condition known as imperforate anus refers to the anus being:
    A. Abnormally wide
    B. Normal in width
    C. Abnormal in location
    D. Abnormally closed   Ref. 2 - p. 727

99. A collection of putty-like or hardened feces in the rectum is known as a:
    A. Fecal collection
    B. Fecal reaction
    C. Fecal impaction
    D. Fecal perforation   Ref. 2 - p. 727

100. The condition of icterus refers to:
    A. Jaundice
    B. Hemorrhage
    C. Anemia
    D. Infection   Ref. 2 - p. 721

101. A normal passage, usually into or through bone is called a:
    A. Foramen
    B. Fissure
    C. Fistula
    D. Sinus   Ref. 2 - p. 572

102. A patient with hypotension has a condition characterized by:
    A. High blood pressure
    B. High temperature
    C. Low temperature
    D. Low blood pressure   Ref. 2 - p. 716

103. A hypophysectomy is an operation performed in which part of the body?:
    A. Lung
    B. Abdomen
    C. Brain
    D. Spine   Ref. 2 - p. 714

104. A hypernephroma is a(an):
    A. Tumor of the kidney
    B. Condition of high blood pressure
    C. Blood clot
    D. Increase in body fat   Ref. 2 - p. 705

105. Emesis refers to:
    A. Shortness of breath
    B. Spitting up blood
    C. A convulsion
    D. Vomiting   Ref. 2 - p. 480

106. A patient who is dyspneic is:
    A. Vomiting
    B. Short of breath
    C. Spitting up blood
    D. Hoarse   Ref. 2 - p. 459

107. Dysphagia refers to difficulty in:
    A. Walking
    B. Breathing
    C. Swallowing
    D. Talking   Ref. 2 - p. 458

108. The injection of air around the kidneys for roentgen diagnosis is called
     perirenal air:
    A. Insufflation
    B. Inspiration
    C. Intubation
    D. Integration   Ref. 2 - p. 746

109. Inflammation of the liver is called:
    A. Hepatomegally
    B. Jaundice
    C. Hepatitis
    D. Cholecystitis   Ref. 2 - p. 667

110. An intrapleural abnormality is located _____ the pleura:
    A. Between
    B. Within
    C. Outside of
    D. Next to   Ref. 2 - p. 753

111. An infraorbital abnormality is located _____ the orbit:
    A. Above
    B. Inside
    C. Within
    D. Beneath   Ref. 2 - p. 740

112. The irrigation or washing out of an organ, such as the stomach, is called:
    A. Rinsing
    B. Salivation
    C. Lavage
    D. Hydrophobia   Ref. 2 - p. 800

113. A _____ is the protrusion of a portion of an organ through a normal
     or abnormal opening:
    A. Haustra
    B. Hernia
    C. Fenestra
    D. Aneurysm   Ref. 2 - p. 620

114. Deficiency of blood in an organ due to partial or complete occlusion of an
     artery is called:
    A. Anemia
    B. Hyperemia
    C. Ischemia
    D. Hypotonia   Ref. 2 - p. 760

115. A tumor made up of fat cells is called a(an):
      A. Adenoma
      B. Meningioma
      C. Lipoma
      D. Lymphoma   Ref. 2 - p. 842

116. A megacolon is a colon which is abnormally:
      A. Small
      B. Large
      C. Located
      D. Twisted   Ref. 2 - p. 888

117. The passage of dark, tarry stool, secondary to bleeding in the gastro-
      intestinal system is known as:
      A. Melena
      B. Melanoma
      C. Hemolysis
      D. Hemophilia   Ref. 2 - p. 891

118. A tumor originating in the membranes covering the brain and spinal
      cord is a:
      A. Hypernephroma
      B. Cancer
      C. Hepatoma
      D. Meningioma   Ref. 2 - p. 896

119. Abnormal smallness of the head is called:
      A. Micrococcus
      B. Micrology
      C. Microcephalus
      D. Hydrocephalus   Ref. 2 - p. 924

120. Migraine is a condition characterized mainly by severe:
      A. Paralysis
      B. Convulsions
      C. Headaches
      D. Blisters   Ref. 2 - p. 929

121. A monarticular abnormality affects:
      A. Many joints
      B. One joint
      C. One ear
      D. One eye   Ref. 2 - p. 938

122. Paralysis of the legs and lower part of the body is known as:
      A. Paraplegia
      B. Poliomyelitis
      C. Ataxia
      D. Hemiatrophy   Ref. 2 - p. 1099

123. A papilloma is a type of:
      A. Infection
      B. Bacterium
      C. X-ray film changer
      D. Tumor   Ref. 2 - p. 1090

124. Bell's palsy is a type of paralysis affecting the:
      A. Arm and hand
      B. Leg and foot
      C. Face
      D. Rib cage   Ref. 2 - p. 1085

125.   A rapid action of the heart which is felt by the patient is called a:
       A.   Fibrillation
       B.   Palpitation
       C.   Impulse
       D.   Cardiogram   Ref. 2 - p. 1084

126.   The slipping of either part of a fractured bone past the other is called:
       A.   Overreaction
       B.   Reduction
       C.   Overgrowth
       D.   Overriding   Ref. 2 - p. 1075

127.   Decrease or loss of bowel motility is referred to as:
       A.   Mechanical ileus
       B.   Meconium ileus
       C.   Dynamic ileus
       D.   Adynamic ileus   Ref. 2 - p. 724

128.   Otitis is an inflammation of the:
       A.   Eye
       B.   Nose and throat
       C.   Ear
       D.   Joints   Ref. 2 - p. 1073

129.   The surgical cutting of bone is referred to as an:
       A.   Osteomalacia
       B.   Osteotomy
       C.   Osteosarcoma
       D.   Ostium   Ref. 2 - p. 1071

130.   A condition marked by softening of the bones is called:
       A.   Osteomyelitis
       B.   Osteochondroma
       C.   Osteomalacia
       D.   Osteosarcoma   Ref. 2 - p. 1070

131.   Chronic degenerative disease of the joints is called:
       A.   Osteoarthritis
       B.   Osteoporosis
       C.   Osteocytosis
       D.   Osteoclasis   Ref. 2 - p. 1068

132.   Surgical incision through the abdomen is called:
       A.   Laparotomy
       B.   Cystostomy
       C.   Paracentesis
       D.   Thoracentesis   Ref. 2 - p. 797

133.   Orchitis refers to inflammation of the:
       A.   Ovary
       B.   Uterus
       C.   Testis
       D.   Orbit   Ref. 2 - p. 1058

134.   The term olfactory refers to the sense of:
       A.   Taste
       B.   Hearing
       C.   Touch
       D.   Smell   Ref. 2 - p. 1040

135. That branch of medicine dealing with the management of pregnancy is called:
   A. Obstetrics
   B. Ophthalmology
   C. Fetometry
   D. Pelvimetry   Ref. 2 - p. 1031

136. A patient who is normotensive has a normal:
   A. Temperature
   B. Blood pressure
   C. Pulse rate
   D. Blood sugar   Ref. 2 - p. 1020

137. Failure of the ends of a fracture to heal and unite is called:
   A. Dislocation
   B. Osteotomy
   C. Nonunion
   D. Malocclusion   Ref. 2 - p. 1019

138. Inflammation of the covering of the brain and spinal cord is called:
   A Meningitis
   B. Encephalitis
   C. Pericarditis
   D. Synovitis   Ref. 2 - p. 896

139. The prefix peri - means:
   A. Beneath
   B. On top of
   C. Inside of
   D. Around   Ref. 2 - p. 1124

140. The abbreviation P.O. stands for:
   A. By mouth (orally)
   B. No liquids allowed
   C. No food allowed
   D. Allergy to drugs   Ref. 2 - p. 1187

141. A small, pinpoint hemorrhage on the skin or mucous membranes is called a(an):
   A. Pessary
   B. Petechia
   C. Hematoma
   D. Ecchymosis   Ref. 2 - p. 1135

142. Inflammation of a vein is known as:
   A. Varicosities
   B. Arteriosclerosis
   C. Phlebitis
   D. Phlebolith   Ref. 2 - p. 1146

143. The science which deals with the functions of the living organism and its parts is:
   A. Pathology
   B. Histology
   C. Anatomy
   D. Physiology   Ref. 2 - p. 1157

144. A medicine that hastens and increases the evacuation from the bowels is
      called a(an):
      A. Cathartic
      B. Antibiotic
      C. Catheter
      D. Antispasmodic   Ref. 2 - p. 264

145. The condition of hardening of the arteries is called:
      A. Arteriovenous
      B. Arteriosclerosis
      C. Arteriography
      D. Arteriology   Ref. 2 - p. 142

146. A network of nerves, or blood or lymph vessels is referred to as a(an):
      A. Tumor
      B. Tubule
      C. Plexus
      D. Anastomosis   Ref. 2 - p. 1175

147. An inflammation involving several joints together is called:
      A. Polyarthritis
      B. Polycythemia
      C. Rheumatism
      D. Hemarthrosis   Ref. 2 - p. 1195

148. A protruding growth or tumor arising from mucous membrane is called a:
      A. Sarcoma
      B. Leukemia
      C. Hemorrhoid
      D. Polyp   Ref. 2 - p. 1197

149. A postprandial symptom is one that occurs:
      A. After meals
      B. Before meals
      C. In the mornings
      D. At bedtime   Ref. 2 - p. 1207

150. A preaortic lymph node lies _____ the aorta:
      A. Behind
      B. Opposite
      C. In front of
      D. Inside of   Ref. 2 - p. 1211

151. A woman who is pregnant for the first time is said to be:
      A. Multigravid
      B. Primigravid
      C. Multiparous
      D. Precocious   Ref. 2 - p. 1218

152. An instrument used to visualize and inspect the rectum is a:
      A. Gastroscope
      B. Telescope
      C. Microscope
      D. Proctoscope   Ref. 2 - p. 1223

153. The falling down or protruding of a viscus from its normal location is
      called:
      A. Prolapse
      B. Volvulus
      C. Distraction
      D. Transposition   Ref. 2 - p. 1225

154. The prognosis of a disease refers to its:
   A. Site of origin
   B. Probable course and outcome
   C. Diagnosis by X-ray studies
   D. Causative agent   Ref. 2 - p. 1225

155. The prefix pseudo-, when used with a word means:
   A. True
   B. Right
   C. Left
   D. False   Ref. 2 - p. 1237

156. A pyarthrosis is a joint cavity containing:
   A. Blood
   B. Air
   C. Pus
   D. Cartilage   Ref. 2 - p. 1253

157. A pyloroplasty is a type of operation performed on the:
   A. Stomach
   B. Kidney
   C. Pelvis
   D. Gall bladder   Ref. 2 - p. 1255

158. Radiculitis is an inflammation of the:
   A. Spinal nerve roots
   B. Brain
   C. Bladder
   D. Rectum   Ref. 2 - p. 1264

159. A skin reaction resulting from exposure to excessive amounts of ionizing radiation is called:
   A. Radioactivity
   B. Radiodermatitis
   C. Psoriasis
   D. Hives   Ref. 2 - p. 1265

160. Destruction of body tissues as a result of radiation exposure is called:
   A. Radiomimetic
   B. Radioactivity
   C. Radiosensitivity
   D. Radionecrosis   Ref. 2 - p. 1266

161. The treatment of disease by roentgen rays or radiant energy is known as:
   A. Radiolucency
   B. Radioresponse
   C. Radiotherapy
   D. Radiometry   Ref. 2 - p. 1266

162. A disease not responsive to treatment is said to be:
   A. Refractory
   B. Curable
   C. Susceptible
   D. Asymptomatic   Ref. 2 - p. 1303

163. A retroperitoneal tumor is located _____ the peritoneum:
   A. In front of
   B. Inside of
   C. On top of
   D. Behind   Ref. 2 - p. 1317

164. Ridges or wrinkles of mucous membrane, such as in the stomach, are called:
   A. Rosettes
   B. Rugae
   C. Roughage
   D. Reticulation   Ref. 2 - p. 1333

165. The salt solution used in intravenous infusions is known as:
   A. Glucose
   B. Epinephrine
   C. Benadryl
   D. Saline   Ref. 2 - p. 1338

166. A sanguinous discharge is one that is:
   A. Bloody
   B. Clear
   C. Cloudy
   D. Thick   Ref. 2 - p. 1341

167. A sarcoma is a type of:
   A. Benign tumor
   B. Infection
   C. Malignant tumor
   D. Surgical instrument   Ref. 2 - p. 1343

168. Spread of malignant cells from one part of the body to the other is called:
   A. Sepsis
   B. Pyrexia
   C. Metastasis
   D. Septicemia   Ref. 2 - p. 909

169. The term nephritis refers to the inflammation of the:
   A. Bladder
   B. Adrenal gland
   C. Brain
   D. Kidney   Ref. 2 - p. 988

170. Ophthalmology is a medical specialty concerned with diseases of the:
   A. Eye
   B. Braine
   C. Ears
   D. Throat   Ref. 2 - p. 1054

171. Osteomyelitis is inflammation of the:
   A. Skin
   B. Bone
   C. Heart
   D. Ear   Ref. 2 - p. 1070

172. The fibrous covering of the heart is known as:
   A. Pericardium
   B. Periosteum
   C. Peritoneum
   D. Epicardium   Ref. 2 - p. 1125

173. The peritoneum is the lining of the walls of the:
   A. Stomach
   B. Brain
   C. Abdomen
   D. Thorax   Ref. 2 - p. 1132

174. Body fluid containing pus is called:
    A. Serous
    B. Purulent
    C. Mucoid
    D. Sanguineous   Ref. 2 - p. 1253

175. _____ bacteria readily cause infection:
    A. Virulent
    B. Motile
    C. Malignant
    D. Encapsulated   Ref. 2 - p. 1692

FOR EACH OF THE FOLLOWING MULTIPLE CHOICE QUESTIONS
SELECT THE ONE MOST APPROPRIATE ANSWER:

176. A patient who is standing, facing front, with the palms of the hands
      turned to the front, is said to be in the:
      A. Lordotic position
      B. Lateral position
      C. Anatomical position
      D. Neutral position   Ref. 3 - p. 11

177. The line that divides the body into two equal parts is the:
      A. Median line
      B. Vertical line
      C. Longitudinal line
      D. Ileopectineal line   Ref. 3 - p. 11

178. That which originates outside an organ, such as a tumor, is said to be:
      A. Intrinsic
      B. Extrinsic
      C. Eccentric
      D. Inherent   Ref. 3 - p. 12

179. Each cell of the body is composed of:
      A. Granules
      B. Fibrils
      C. Chromosomes
      D. Protoplasm   Ref. 3 - p. 16

180. The centrally placed seat of activity of each cell is known as the:
      A. Nucleus
      B. Cytoplasm
      C. Centrosome
      D. Centromere   Ref. 3 - p. 16

181. Growth and repair of body tissues take place by:
      A. Cell elongation
      B. Maturation
      C. Cell division
      D. Replacement   Ref. 3 - p. 16

182. The process of cell splitting in which chromosomes are formed and
      migrate to opposite poles, is called:
      A. Multiplication
      B. Reproduction
      C. Cytology
      D. Mitosis   Ref. 3 - p. 16

183. A group of cells similar in both form and function is called:
      A. Organs
      B. Tissue
      C. System
      D. Colloid   Ref. 3 - p. 17

184. Membranes covering surfaces and lining tubes and cavities:
      A. Ectoplasm
      B. Endoderm
      C. Epiphyseal
      D. Epithelium   Ref. 3 - p. 17

185. Membranes which line joint cavities and bursae:
      A. Areolar
      B. Synovial
      C. Syncytial
      D. Cuboidal   Ref. 3 - p. 17

186. The function of the connective or areolar tissues is:
      A. Circulation
      B. Transmission
      C. Secretion
      D. Support   Ref. 3 - p. 17

187. _____ plane divides the body into two equal parts from front to
      back through the sagittal suture of the skull:
      A. Coronal
      B. Frontal
      C. Median
      D. Lateral   Ref. 3 - p. 11

188. Skeletal muscle is:
      A. Involuntary
      B. Voluntary
      C. Visceral
      D. None of the above   Ref. 3 - p. 11

189. Visceral muscle is found in the:
      A. Walls of many organs
      B. Extremities
      C. Heart
      D. Ligaments   Ref. 3 - p. 17

190. The active, functioning tissue in an organ is the:
      A. Stroma
      B. Cytoplasm
      C. Mesothelium
      D. Parenchyma   Ref. 3 - p. 17

191. The heart, blood vessels and lymphatics are known, collectively, as the:
      A. Circulatory system
      B. Digestive system
      C. Nervous system
      D. Skeletal system   Ref. 3 - p. 19

192. The ductless glands are part of the:
      A. Reproductive system
      B. Endocrine system
      C. Glandular system
      D. Urinary system   Ref. 3 - p. 19

193. The air passages and lungs comprise the:
      A. Muscular system
      B. Reproductive system
      C. Respiratory system
      D. Circulatory system   Ref. 3 - p. 19

194. The kidneys, ureters, bladder and urethra make up the:
      A. Reproductive system
      B. Urinary system
      C. Endocrine system
      D. Genital system   Ref. 3 - p. 19

195. The body cavity is divided into the:
      A. Thorax and abdomen
      B. Abdomen, pelvis and cranium
      C. Abdomen and cranium
      D. Thorax, abdomen and pelvis   Ref. 3 - p. 19

196. The female reproductive cell is the:
      A. Uterus
      B. Fallopian tube
      C. Ovary
      D. Ovum   Ref. 3 - p. 19

197. The spermatozoa are formed in the:
      A. Testis
      B. Adrenal
      C. Pancreas
      D. Ovary   Ref. 3 - p. 20

198. The unborn young in the human is called:
      A. Ovum
      B. Fetus
      C. Chromosome
      D. Premature infant   Ref. 3 - p. 20

199. _____ plane divides the body vertically from side to side:
      A. Coronal
      B. Median
      C. Lateral
      D. Horizontal   Ref. 3 - p. 11

200. When an organ is found to be on the side of the body opposite to its
      normal placement, _____ is said to be present:
      A. Dislocation
      B. Transposition
      C. Displacement
      D. Dislodgement   Ref. 3 - p. 21

201. The outer layer of the skin is the:
      A. Epithelium
      B. Desquamation
      C. Epidermis
      D. Dermis   Ref. 3 - p. 25

202. The degree of darkness of the skin is determined by the
      A. Sensitivity
      B. Granules
      C. Thickness
      D. Pigment   Ref. 3 - p. 25

203. The horny layer of the skin is thicker on what part of the body?:
      A. Palms and soles
      B. Face
      C. Abdomen
      D. Scalp   Ref. 3 - p. 25

204.   The layer of connective tissue lying between the skin and the muscles
       is the:
       A. Mesoderm
       B. Ligament
       C. Muscular membrane
       D. Subcutaneous tissue   Ref. 3 - p. 26

205.   The nails are a modification of what structure?:
       A. Hair
       B. Skin
       C. Bone
       D. Cartilage   Ref. 3 - p. 26

206.   That portion of the hair lying below the skin surface is the:
       A. Root
       B. Shaft
       C. Follicle
       D. Bulb   Ref. 3 - p. 26

207.   The small glands which secrete an oily substance to lubricate the hair
       are the:
       A. Sweat glands
       B. Ductless glands
       C. Sebaceous glands
       D. Apocrine glands   Ref. 3 - p. 26

208.   The sweat glands help to:
       A. Dissipate body heat
       B. Produce body heat
       C. Cleanse the skin
       D. All of the above   Ref. 3 - p. 26

209.   The first reaction of the skin to external radiation is:
       A. Hair loss
       B. Blisterine
       C. Ulceration
       D. Reddening   Ref. 3 - p. 28

210.   A small bone is known as a(an):
       A. Ostium
       B. Ossicle
       C. Condyle
       D. Eminence   Ref. 3 - p. 29

211.   A rounded, knob-like projection of bone is called a(an):
       A. Epicondyle
       B. Ala
       C. Condyle
       D. Sulcus   Ref. 3 - p. 29

212.   In human anatomy dorsal is synonomous with:
       A. Anterior
       B. Lateral
       C. Superior
       D. Posterior   Ref. 3 - p. 12

213.   The portion of a structure close to its origin is called:
       A. Distal
       B. Proximal
       C. Caudal
       D. Volar   Ref. 3 - p. 12

214. To bring towards the median plane is to:
    A. Adduct
    B. Abduct
    C. Circumduct
    D. Rotate

Ref. 3 - p. 13

215. The distal end of a short bone is referred to as its:
    A. Neck
    B. Base
    C. Shaft
    D. Head

Ref. 3 - p. 31

216. When lying face down a patient is in the _____ position:
    A. Supine
    B. Prone
    C. Decubitus
    D. Extended

Ref. 3 - p. 13

217. The dense, compact portion of a bone is the:
    A. Medulla
    B. Metaphysis
    C. Epiphysis
    D. Cortex

Ref. 3 - p. 31

218. The digestive system is also known as the:
    A. Nutritional system
    B. Endocrine system
    C. Alimentary tract
    D. None of the above

Ref. 3 - p. 19

219. The central cavity in the shaft of a long bone is the:
    A. Cortex
    B. Foramen
    C. Medulla
    D. Fossa

Ref. 3 - p. 31

220. The layer of cancellous bone between the inner and outer tables of the skull is the:
    A. Medulla
    B. Marrow cavity
    C. Diaphysis
    D. Diploe

Ref. 3 - p. 31

221. That part of a bone developing from a primary ossification center and forming the shaft of the bone is known as the:
    A. Metaphysis
    B. Diaphysis
    C. Apophysis
    D. Epiphysis

Ref. 3 - p. 32

222. The secondary ossification center at the end of bone is called the:
    A. Metaphysis
    B. Diaphysis
    C. Apophysis
    D. Epiphysis

Ref. 3 - p. 32

223. The X-ray examination for bone age in a child depends upon the appearance time of what structures?:
A. Primary ossification centers
B. Epiphyses
C. Metacarpals
D. Radius and ulna   Ref. 3 - p. 32

224. The radiolucent lines visible at the ends of the long bones in children actually represent:
A. Bone gaps
B. Fibrous tissue
C. Cartilage
D. Fat   Ref. 3 - p. 33

225. The bones of the shoulder girdle include the:
A. Clavicle and sternum
B. Scapula and dorsal spine
C. Humerus and ulna
D. Scapula and clavicle   Ref. 3 - p. 34

226. The navicular bone is located in the:
A. Skull
B. Proximal carpal row
C. Distal carpal row
D. Pelvis   Ref. 3 - p. 34

227. The connection of the scapula with the bones of the trunk is at the:
A. Acromio-clavicular joint
B. Glenoid-humeral joint
C. Vertebral margin
D. Posterior chest wall   Ref. 3 - p. 35

228. The outer border of the scapula is termed the:
A. Vertebral
B. Superior
C. Inferior
D. Axillary   Ref. 3 - p. 35

229. The bony ridge which passes across the posterior border of the scapula is the:
A. Neck
B. Angle
C. Spine
D. Acromion   Ref. 3 - p. 36

230. The bony projection of the scapula that can be felt through the skin about the shoulder is the:
A. Spine
B. Coracoid
C. Notch
D. Acromion   Ref. 3 - p. 36

231. The depression or cavity of the scapula that articulates with the head of the humerus is called the:
A. Acetabulum
B. Glenoid
C. Condyle
D. Foramen   Ref. 3 - p. 36

232. The fertilized ovum up to the third month of development is called:
      A. Zygote
      B. Fetus
      C. Ovary
      D. Embryo   Ref. 3 - p. 20

233. The inner (medial) end of the clavicle articulates with which structure?:
      A. Ribs
      B. Scapula
      C. Humerus
      D. Sternum   Ref. 3 - p. 36

234. The depression in the skin above the clavicle is called the:
      A. Suprasternal notch
      B. Supraspinatous fossa
      C. Supraclavicular fossa
      D. Infraclavicular fossa   Ref. 3 - p. 36

235. The rounded, proximal articular end of the humerus is the:
      A. Head
      B. Neck
      C. Condyle
      D. Spine   Ref. 3 - p. 37

236. What is the name of the bony prominence found on the upper, outer
      surface of the humerus?:
      A. Epicondyle
      B. Trochanter
      C. Greater tubercle
      D. Malleolus   Ref. 3 - p. 37

237. That portion of the distal humerus that articulates with the ulna is the:
      A. Trochlea
      B. Condyle
      C. Styloid
      D. Patella   Ref. 3 - p. 37

238. The flat, circular end of the proximal radius that articulates with the
      humerus is called the:
      A. Neck
      B. Head
      C. Trochanter
      D. Shaft   Ref. 3 - p. 37

239. The bony prominence of the distal radius at the wrist joint is the:
      A. Tuberosity
      B. Epicondyle
      C. Styloid process
      D. Malleolus   Ref. 3 - p. 40

240. The founded, proximal end of the ulna which forms the tip of the elbow
      is the:
      A. Capitulum
      B. Trochlea
      C. Coronoid
      D. Olecranon   Ref. 3 - p. 40

241.   What is the name of the carpal bone that is superimposed on the
        triquetral bone in an A P projection of the wrist?:
        A. Navicular
        B. Pisiform
        C. Lunate
        D. Hamate   Ref. 3 - p. 41

242.   What is the name of the wrist bone that has a hook-like process?:
        A. Navicular
        B. Lesser multangular
        C. Hemate
        D. Triquetral   Ref. 3 - p. 41

243.   What is the name of the wrist bone that articulates with the first
        metacarpal?:
        A. Navicular
        B. Hemate
        C. Lesser multangular
        D. Greater multangular   Ref. 3 - p. 41

244.   The navicular bone does not come in contact with which one of the
        following?:
        A. Greater multangular
        B. Lunate
        C. Capitate
        D. Triquetral   Ref. 3 - p. 41

245.   The thumb has how many phalanges?:
        A. One
        B. Two
        C. Three
        D. Four   Ref. 3 - p. 41

246.   Elevation and deformity of the scapula is known as:
        A. Sprengel's deformity
        B. Madelung's deformity
        C. Pott's disease
        D. Kyphosis   Ref. 3 - p. 42

247.   What is the name of the prominent projection which can be felt at the
        medial margin of the elbow?:
        A. Olecranon process
        B. Lateral epicondyle
        C. Styloid process
        D. Medial epicondyle   Ref. 3 - p. 42

248.   Which of the following bony prominences can not be felt through the skin?:
        A. Olecranon process
        B. Humeral epicondyle
        C. Deltoid tubercle
        D. Acromion process   Ref. 3 - p. 42

249.   The only wrist bone that can be felt through the skin as a separate
        structure is the:
        A. Navicular
        B. Pisiform
        C. Hamate
        D. Lunate   Ref. 3 - p. 42

250.   The "knuckles" are really what anatomic structures?:
        A. Interphalangeal joints
        B. Carpo-metacarpal joints
        C. Intercarpal joints
        D. Metacarpo-phalangeal joints   Ref. 3 - p. 42

251.   When a patient is lying down with the palms up (supinated) the radius
        lies _____ the ulna:
        A. Medial to
        B. Below
        C. Behind
        D. Lateral to   Ref. 3 - p. 43

252.   All of the digitis except the thumb have how many interphalangeal joints?:
        A. One
        B. Two
        C. Three
        D. Four   Ref. 3 - p. 42

253.   The dermis is derived from embryonic:
        A. Mesoderm
        B. Ectoderm
        C. Endoderm
        D. Epidermis   Ref. 3 - p. 26

254.   Hair is derived from the _____ layer of the skin:
        A. Germinal
        B. Granular
        C.ᐟ Horny
        D. Translucent   Ref. 3 - p. 26

255.   The cavity in the innominate bone that articulates with the femoral
        head is the:
        A. Glenoid
        B. Acetabulum
        C. Sacro-iliac joint
        D. Semilunar notch   Ref. 3 - p. 46

256.   The upper, curved border of the ilium that can be felt through the skin
        of the lateral abdominal wall is the:
        A. Notch
        B. Crest
        C. Tuberosity
        D. Spine   Ref. 3 - p. 47

257.   The prominent pointed projection of the ilium that can be felt through the
        skin and is a useful landmark in positioning, is the:
        A. Posterior spine
        B. Anterior superior spine
        C. Tuberosity
        D. Coccyx   Ref. 3 - p. 47

258.   The large opening in the lower innominate bone that is bounded by the
        ischium and pubis is called the:
        A. Jugular foramen
        B. Foramen magnum
        C. Obturator foramen
        D. Foramen of Winslow   Ref. 3 - p. 46

259. The large, rounded prominence of the ischium which supports the body weight while in the sitting position is the:
    A. Tuberosity
    B. Tubercle
    C. Ramus
    D. Spine   Ref. 3 - p. 48

260. The pointed process of the ischium which is directed medialward and whose size is important during childbirth is the:
    A. Tuberosity
    B. Tubercle
    C. Ramus
    D. Spine   Ref. 3 - p. 48

261. The joint that separates the two pubic bones is the:
    A. Syndrome
    B. Symphysis
    C. Sacro-iliac
    D. Ischial   Ref. 3 - p. 48

262. The lower anterior part of the innominate bone is formed by the:
    A. Iliac wing
    B. Pubis
    C. Ischium
    D. Sacrosciatic notch   Ref. 3 - p. 48

263. The artificial upper border of the true pelvis is called the:
    A. Symphysis
    B. Posterior sagittal
    C. Inlet
    D. Outlet   Ref. 3 - p. 48

264. The lower opening of the pelvic cavity is the:
    A. Arch
    B. Tubercle
    C. Inlet
    D. Outlet   Ref. 3 - p. 48

265. The posterior boundary of the true pelvis is the:
    A. Ischium
    B. Sacrum and coccyx
    C. Ilium
    D. Pubis   Ref. 3 - p. 48

266. During childbirth, the fetal head must pass through the:
    A. Obturator foramen
    B. Sciatic notch
    C. Pelvic inlet and outlet
    D. Diaphragm   Ref. 3 - p. 48

267. The false pelvis is really part of the:
    A. Thorax
    B. Iliac bone
    C. Abdomen
    D. Pubis and ischium   Ref. 3 - p. 48

268. That part of the femur which connects the head with the shaft is the:
    A. Neck
    B. Crest
    C. Body
    D. Condyle   Ref. 3 - p. 50

269.   The large, palpable prominence on the outer (lateral) border of the
       femur is the:
       A.   Lesser trochanter
       B.   Greater trochanter
       C.   Greater tuberosity
       D.   Lesser multangular   Ref. 3 - p. 50

270.   The small, round process on the upper, medial border of the femoral
       shaft is the:
       A.   Greater tuberosity
       B.   Lesser tuberosity
       D.   Greater trochanter
       D.   Lesser trochanter   Ref. 3 - p. 50

271.   The oblique bony ridge on the proximal femoral shaft is called the:
       A.   Medial condyle
       B.   Iliac crest
       C.   Intertrochanteric crest
       D.   Epicondyle   Ref. 3 - p. 50

272.   The large knob-like process on the medial half of the distal femur is the:
       A.   Greater trochanter
       B.   Medial condyle
       C.   Medial tuberosity
       D.   Medial tubercle   Ref. 3 - p. 50

273.   The deep notch separating the medial and lateral femoral condyles is the:
       A.   Intertrochanteric crest
       B.   Semilunar notch
       C.   Intercondyloid fossa
       D.   Glenoid fossa   Ref. 3 - p. 50

274.   The large oval bone that lies within the tendon of the quadriceps
       muscle is the:
       A.   Patella
       B.   Fabella
       C.   Tibia
       D.   Calcaneus   Ref. 3 - p. 50

275.   Sweat glands are most numerous on the:
       A.   Axilla and groin
       B.   Scalp and forehead
       C.   Chest and back
       D.   Palms and soles   Ref. 3 - p. 27

276.   The round prominence on the anterior surface of the proximal tibia is the:
       A.   Tibial spine
       B.   Tibial tuberosity
       C.   Articular facet
       D.   Condyle   Ref. 3 - p. 51

277.   The large prominence of the distal tibia that forms part of the inner
       (medial) border of the ankle is the:
       A.   Medial condyle
       B.   Lateral condyle
       C.   Lateral malleolus
       D.   Medial malleolus   Ref. 3 - p. 51

278. The upper, bulbous end of the fibula is the:
      A. Neck
      B. Head
      C. Malleolus
      D. Shaft   Ref. 3 - p. 51

279. The lower prominence of the fibula that forms part of the outer border
      of the ankle is the:
      A. Medial malleolus
      B. Lateral malleolus
      C. Medial condyle
      D. Lateral condyle   Ref. 3 - p. 51

280. The medial and lateral malleoli form the:
      A. Ankle mortise
      B. Subtalar joint
      C. Knee joint
      D. Intertarsal joint   Ref. 3 - p. 51

281. After radiation injury skin is regenerated from:
      A. Subcutaneous fat
      B. Granular layer and sebaceous glands
      C. Germinal layer and hair follicles
      D. Reticular layer   Ref. 3 - p. 28

282. The talus articulates below with which bone?:
      A. Calcaneus
      B. Cuneiform
      C. Tibia
      D. Metatarsal   Ref. 3 - p. 51

283. The talus articulates in front with which bone?:
      A. Calcaneus
      B. Cuneiform
      C. Navicular
      D. Tibia   Ref. 3 - p. 51

284. The three ankle bones that lie side by side behind the metatarsal bones
      are the:
      A. Cuboids
      B. Cuneiforms
      C. Metacarpals
      D. Phalanges   Ref. 3 - p. 53

285. The cube-shaped bone lying in front of the calcaneus is the:
      A. Navicular
      B. Sustentaculum
      C. Cuboid
      D. Oscalsis   Ref. 3 - p. 53

286. The bones that form the instep of the foot are the:
      A. Cuneiforms
      B. Phalanges
      C. Metacarpals
      D. Metatarsals   Ref. 3 - p. 53

287. In the foot, there are _____ phalanges:
      A. 7
      B. 5
      C. 14
      D. 2   Ref. 3 - p. 53

288.   The longitudinal arch is found on the:
        A. Dorsum of the foot
        B. Plantar surface of the foot
        C. Palmar surface of the hand
        D. Back of the wrist   Ref. 3 - p. 54

289.   When the patella remains as two or more ununited, separate bony centers,
        it is said to be:
        A. Bifurcated
        B. Biphasic
        C. Bipartite
        D. Bidirectional   Ref. 3 - p. 54

290.   The sacro-iliac joints are directed _____ from front to back:
        A. At 45 degrees
        B. Horizontally
        C. Transversely
        D. Obliquely   Ref. 3 - p. 55

291.   The ankle joint lies about _____ the tips of the malleoli:
        A. 3/4 inch below
        B. 3 inches above
        C. 3/4 inch above
        D. 3 inches below   Ref. 3 - p. 56

292.   The number of cervical vertebra is:
        A. 5
        B. 7
        C. 10
        D. 12   Ref. 3 - p. 60

293.   The number of dorsal vertebra is:
        A. 5
        B. 7
        C. 10
        D. 12   Ref. 3 - p. 60

294.   The number of lumbar vertebra is:
        A. 5
        B. 7
        C. 10
        D. 12   Ref. 3 - p. 60

295.   All animals, including humans, who have a vertebral column are known as
        A. Invertebrates
        B. Amphibians
        C. Vertebrates
        D. Primates   Ref. 3 - p. 60

296.   How many vertebrae become fused to form the sacrum?:
        A. 3
        B. 5
        C. 7
        D. 9   Ref. 3 - p. 60

297.   What type of natural curve is found in the cervical spine?:
        A. Scoliosis
        B. Kyphosis
        C. Lordosis
        D. Kypho-scoliosis   Ref. 3 - p. 60

298.   What type of natural curve is found in the dorsal spine?:
       A. Scoliosis
       B. Kyphosis
       C. Lordosis
       D. Kyphoscoliosis   Ref. 3 - p. 60

299.   The natural curve of the lumbar spine is similar to the curve found in the:
       A. Cervical spine
       B. Dorsal spine
       C. Sacrum
       D. Coccyx   Ref. 3 - p. 60

300.   The solid, block-shaped portion of a vertebra is called the:
       A. Lamina
       B. Pedicle
       C. Body
       D. Disc space   Ref. 3 - p. 60

301.   The two laminae of a vertebra unite in the midline posteriorly to form the:
       A. Articular facet
       B. End plate
       C. Spinous process
       D. Transverse process   Ref. 3 - p. 61

302.   The opening in the vertebra through which the spinal cord passes is the:
       A. Condylar foramen
       B. Vertebral foramen
       C. Intervertebral foramen
       D. Anterior foramen   Ref. 3 - p. 61

303.   The first cervical vertebra is known as the:
       A. Coccyx
       B. Atlas
       C. Axis
       D. Dens   Ref. 3 - p. 61

304.   The first cervical vertebra is distinctive because it has no:
       A. Body
       B. Posterior arch
       C. Facet
       D. Transverse process   Ref. 3 - p. 61

305.   The second cervical vertebra is termed the:
       A. Atlas
       B. Axis
       C. Dens
       D. Body   Ref. 3 - p. 61

306.   The second cervical vertebra has a large tooth-like process projecting
       from it which fits into the arch of C1. This process is called the:
       A. Atlas
       B. Axis
       C. Dens
       D. Condyle   Ref. 3 - p. 61

307.   Rotation of the head takes place between:
       A. The base of the skull and C1
       B. C1 and C2
       C. C3 and C4
       D. C6 and C7   Ref. 3 - p. 61

308.   Flexion of the head takes place between:
        A. The base of the skull and C1
        B. C1 and C2
        C. C3 and C4
        D. C6 and C7   Ref. 3 - p. 61

309.   The spinous process of which cervical vertebra is the most prominent?:
        A. C1
        B. C2
        C. C5
        D. C7   Ref. 3 - p. 62

310.   Which vertebrae have facets for articulation with the ribs?:
        A. Cervical
        B. Dorsal
        C. Lumbar
        D. Sacral   Ref. 3 - p. 62

311.   The prominent upper anterior edge of the sacrum is called the:
        A. Coccyx
        B. Ala
        C. Promontory
        D. Body   Ref. 3 - p. 62

312.   The sacral nerves extend out through the:
        A. Sacral canal
        B. Sacral foramina
        C. Sacral window
        D. Sacral notch   Ref. 3 - p. 62

313.   The passage that extends through the sacrum from top to bottom is the:
        A. Sacral canal
        B. Sacral foramen
        C. Sacral window
        D. Sacral notch   Ref. 3 - p. 62

314.   The ring-like layer of fibrous and cartilage tissue between the articular
        plates of adjacent vertebrae is the:
        A. Nucleus pulposus
        B. Intervertebral joint
        C. Annulus fibrosus
        D. Centrum   Ref. 3 - p. 63

315.   The pulpy center of the intervertebral disc space is the:
        A. Nucleus pulposus
        B. Intervertebral foramen
        C. Annulus fibrosus
        D. Centrum   Ref. 3 - p. 63

316.   The forward displacement of one vertebra in relation to an adjacent
        vertebra, secondary to a defect in the neural arch, is called:
        A. Dislocation
        B. Spondylolisthesis
        C. Fracture
        D. Rotation   Ref. 3 - p. 63

317.   If the last lumbar vertebra is partly or completely fused with the sacrum,
        it is said to be:
        A. Lumbarized
        B. Occipitalized
        C. A hemivertebra
        D. Sacralized   Ref. 3 - p. 63

318. The suprasternal notch is at the same level as:
    A. C3-C4
    B. C6-C7
    C. D2-D3
    D. D6-D7   Ref. 3 - p. 64

319. The sternal angle is at the same level as:
    A. C3-C4
    B. C5-C6
    C. D4-D5
    D. D10-D11   Ref. 3 - p. 64

320. The umbilicus is at the same level as:
    A. D5-D6
    B. D10-D11
    C. L1-L2
    D. L3-L4   Ref. 3 - p. 64

321. The anterior superior iliac spine is at the same level as:
    A. L1-L2
    B. L4-L5
    C. S2
    D. Coccyx   Ref. 3 - p. 64

322. There are _____ pairs of ribs:
    A. 8
    B. 10
    C. 12
    D. 14   Ref. 3 - p. 67

323. The meaning if sinus is:
    A. A cavity within bone
    B. Passage from body cavity to the outside
    C. Lake-like cavity containing blood
    D. All of the above   Ref. 3 - p. 29

324. The upper segment of the sternum is called the:
    A. Xiphoid
    B. Body
    C. Manubrium
    D. Head   Ref. 3 - p. 67

325. The lowermost, pointed end of the sternum is called the:
    A. Manubrium
    B. Xiphoid
    C. Body
    D. Angle   Ref. 3 - p. 67

326. The concave notch or depression in the upper sternal segment is called the:
    A. Costal notch
    B. Clavicular notch
    C. Suprasternal notch
    D. Xiphoid   Ref. 3 - p. 67

327. The ridge in the upper portion of the sternum which is easily palpated, and lies opposite the second pair of ribs, is called the:
    A. Angle of Petit
    B. Sternal angle
    C. Costal angle
    D. Costovertebral angle   Ref. 3 - p. 68

328.   The upper seven pair of ribs are:
        A.   True ribs
        B.   False ribs
        C.   Floating ribs
        D.   Rudimentary ribs   Ref. 3 - p. 68

329.   The long, flat, curved part of rib is called the:
        A.   Head
        B.   Neck
        C.   Shaft
        D.   Tubercle   Ref. 3 - p. 68

330.   There are _____ pair of costal cartilages:
        A.   6
        B.   10
        C.   12
        D.   14   Ref. 3 - p. 68

331.   The junction of the anterior end of a rib with its cartilage is called the:
        A.   Costovertebral joint
        B.   Costochondral joint
        C.   Sternocostal joint
        D.   Xiphisternal joint   Ref. 3 - p. 68

332.   An extra rib, arising from C7, is called a _____ rib:
        A.   Floating
        B.   True
        C.   Cervical
        D.   Lumbar   Ref. 3 - p. 68

333.   The condition of a depressed sternum ("funnel chest") is known as:
        A.   Pectus excavatum
        B.   Double sternum
        C.   Pectus carinatum
        D.   Flail chest   Ref. 3 - p. 69

334.   The gall bladder is found in what region of the body?:
        A.   Above the diaphragm
        B.   Under the left costal margin
        C.   In the midline
        D.   Under the right costal margin Ref. 3 - p. 69

335.   The spleen lies in which portion of the body?:
        A.   Above the diaphragm
        B.   Under the left costal margin
        C.   In the midline
        D.   Under the right costal margin Ref. 3 - p. 69

336.   The roof, or superior region of the skull is the:
        A.   Floor
        B.   Nasion
        C.   Base
        D.   Vertex   Ref. 3 - p. 73

337.   The type of marrow found in the long bones of adults is:
        A.   Red
        B.   Yellow
        C.   Aplastic
        D.   Hyperplastic   Ref. 3 - p. 31

338. That point on the posterior part of the skull where the two parietal bones meet is termed the:
   A. Nasion
   B. Bregma
   C. Pterion
   D. Lambda   Ref. 3 - p. 73

339. Which of the following skull bones are paired?:
   A. Frontal
   B. Occipital
   C. Parietal
   D. Sphenoid   Ref. 3 - p. 73

340. Which of the following facial bones are single?:
   A. Maxilla
   B. Zygomatic
   C. Nasal
   D. Mandible   Ref. 3 - p. 73

341. The vertical partition between the two nasal cavities is called the:
   A. Concha
   B. Nasal septum
   C. Nasal spine
   D. Canthus   Ref. 3 - p. 75

342. That point at the outer edge of the eye where the upper and lower lids meet is the:
   A. Punctum
   B. Inner canthus
   C. External canthus
   D. Cornea   Ref. 3 - p. 75

343. The skull bone which forms the forehead is the _____ bone:
   A. Temporal
   B. Frontal
   C. Occipital
   D. Parietal   Ref. 3 - p. 75

344. The bony prominences above the orbits which are covered by the eyebrows are the:
   A. Infraorbital foramina
   B. Infraorbital ridges
   C. Supraorbital ridges
   D. Supraorbital foramina   Ref. 3 - p. 75

345. The parietal bones are separated in the midline by the following suture:
   A. Sagittal
   B. Coronal
   C. Lambdoidal
   D. Squamosal   Ref. 3 - p. 75

346. The anterior borders of the parietal bones lie next to which bone?:
   A. Temporal
   B. Nasal
   C. Occipital
   D. Frontal   Ref. 3 - p. 75

347.   The greatest transverse diameter of the skull lies between the two
               _____ bones:
        A.   Temporal
        B.   Parietal
        C.   Nasal
        D.   Zygomatic   Ref. 3 - p. 75

348.   The occipital bone forms which part of the skull?:
        A.   Top
        B.   Side
        C.   Front
        D.   Back   Ref. 3 - p. 75

349.   The occipital and parietal bones are separated by which suture?:
        A.   Lambdoid
        B.   Squamosal
        C.   Coronal
        D.   Sagittal   Ref. 3 - p. 75

350.   The large opening in the base of the skull through which the brain stem
        passes is called the:
        A.   Jugular foramen
        B.   Carotid foramen
        C.   Foramen magnum
        D.   Auditory foramen   Ref. 3 - p. 75

351.   The two bony prominences on the occipital bone that articulate with
        C1 are the:
        A.   Occipital foramina
        B.   Occipital condyles
        C.   Styloid processes
        D.   Mandibular condyles   Ref. 3 - p. 75

352.   The bony prominence of the outer surface of the occipital bone which
        can easily be palpated is called the:
        A.   Internal occipital protuberance
        B.   External occipital protuberance
        C.   Occipital condyle
        D.   Occipital crest   Ref. 3 - p. 75

353.   The large bony prominence of the temporal bone that extends down behind
        the ear is called the:
        A.   Zygomatic
        B.   Petrous
        C.   Mastoid
        D.   Squamosal   Ref. 3 - p. 76

354.   That part of the temporal bone which contains the inner ear and is shaped
        like a pyramid is the:
        A.   Zygomatic
        B.   Petrous
        C.   Mastoid
        D.   Squamosal   Ref. 3 - p. 76

355.   The sharp, pointed process that extends down from the base of the
        temporal bone is the _____ process:
        A.   Zygomatic
        B.   Mastoid
        C.   Alveolar
        D.   Styloid   Ref. 3 - p. 76

356.   The outer opening of the ear canal in the temporal bone is the:
       A. Internal auditory meatus
       B. External auditory meatus
       C. External canthus
       D. Optic foramen   Ref. 3 - p. 76

357.   The membrane separating the ear canal from the middle ear is the:
       A. Tympanic membrane
       B. Diaphragm
       C. Cornea
       D. Dura mater   Ref. 3 - p. 76

358.   Where does the body of the sphenoid bone lie?:
       A. Midline of the floor of the skull
       B. Midline in the roof of the skull
       C. In the back of the skull
       D. In the orbit   Ref. 3 - p. 76

359.   The two pairs of sphenoid wings are termed:
       A. Medial and lateral
       B. Superior and inferior
       C. Greater and lesser
       D. Internal and external   Ref. 3 - p. 76

360.   The opening in the front of the sphenoid bone through which the nerves
       of sight pass, is called the:
       A. Auditory meatus
       B. Auditory canal
       C. Olfactory groove
       D. Optic foramen   Ref. 3 - p. 76

361.   The saddle-shaped depression in the sphenoid bone which contains the
       pituitary gland is known as the:
       A. Middle fossa
       B. Optic foramen
       C. Sella turcica
       D. Orbital fissure   Ref. 3 - p. 76

362.   The posterior wall of the pituitary fossa is called the:
       A. Dorsum sellae
       B. Clivus
       C. Sphenoid ridge
       D. Sella turcica   Ref. 3 - p. 76

363.   The two small bony prominences on top of the posterior wall of the
       pituitary fossa are known as the _____ processes:
       A. Anterior clinoid
       B. Posterior clinoid
       C. Pterygoid
       D. Styloid   Ref. 3 - p. 76

364.   That portion of the ethmoid bone containing numerous small openings
       is the:
       A. Mastoid process
       B. Clivus
       C. Cribiform plate
       D. Paranasal sinus   Ref. 3 - p. 76

365.   The nerves which pass through the openings in the ethmoid bone transmit
        the sense of:
        A. Sight
        B. Hearing
        C. Taste
        D. Smell   Ref. 3 - p. 76

366.   The flat, pointed bony projection that extends in the midline for a short
        distance above the ethmoid bone is known as the:
        A. Styloid process
        B. Cribiform plate
        C. Crista galli
        D. Sella turcica   Ref. 3 - p. 76

367.   The maxilla comprises the:
        A. Orbits
        B. Nasal fossa
        C. Lower jaw
        D. Upper jaw   Ref. 3 - p. 78

368.   The process of the maxillary bone containing the teeth is the _____
        process:
        A. Clinoid
        B. Alveolar
        C. Styloid
        D. Pterygoid   Ref. 3 - p. 78

369.   The small bony projection located just below the nasal septum is the:
        A. Palatine process
        B. Anterior nasal spine
        C. Styloid process
        D. Cribiform plate   Ref. 3 - p. 78

370.   That portion of the maxilla that forms part of the hard palate is the:
        A. Alveolar process
        B. Uvula
        C. Palatine process
        D. Tonsil   Ref. 3 - p. 78

371.   The bones which form the prominent part of the cheek are the:
        A. Zygomatic
        B. Nasal
        C. Vomer
        D. Lacrimal   Ref. 3 - p. 78

372.   The spinatus muscles insert in the _____ of the humerus:
        A. Lesser tubercle
        B. Greater tubercle
        C. Surgical neck
        D. Deltoid tubercle   Ref. 3 - p. 37

373.   The movable part of the nose is composed of:
        A. Bone
        B. Tendon
        C. Cartilage
        D. Muscle   Ref. 3 - p. 78

374. The small, flat bones which contain a groove for tear duct are called the _____ bones:
    A. Zygomatic
    B. Lacrimal
    C. Palatine
    D. Vomer   Ref. 3 - p. 78

375. The thin, curved bone that extends into the nasal cavity, forming an incomplete shelf, is called the:
    A. Inferior concha
    B. Nasal septum
    C. Nasal spine
    D. Vomer   Ref. 3 - p. 80

376. The single flat bone lying below and behind the ethmoid, shaped like a "ploughshare," is the:
    A. Zygoma
    B. Lacrimal
    C. Concha
    D. Vomer   Ref. 3 - p. 80

377. The junction of the vertical and horizontal portion of the mandible is called the:
    A. Ramus
    B. Condyle
    C. Angle
    D. Symphysis   Ref. 3 - p. 80

378. The horizontal portion of the mandible is termed the:
    A. Ramus
    B. Angle
    C. Neck
    D. Body   Ref. 3 - p. 89

379. The vertical portion of the mandible is the:
    A. Ramus
    B. Angle
    C. Body
    D. Symphysis   Ref. 3 - p. 80

380. The midline portion of the mandible where its two halves have united is the:
    A. Angle
    B. Symphysis
    C. Mental foramen
    D. Mental protuberance   Ref. 3 - p. 80

381. The prominence at the lower part of the mandible, which forms the chin, is the:
    A. Zygomatic arch
    B. Mandibular condyle
    C. Maxillary tuberosity
    D. Mental protuberance   Ref. 3 - p. 80

382. The small opening in the horizontal portion of the mandible, through which passes the mandibular nerve, is called the:
    A. Mental foramen
    B. Mandibular fossa
    C. ' Mandibular notch
    D. Alveolar ridge   Ref. 3 - p. 81

383.   That portion of the mandible forming part of the temporomandibular
        joint is called the:
        A. Capitulum
        B. Neck
        C. Ramus
        D. Body   Ref. 3 - p. 81

384.   The horseshoe-shaped bone lying in front of the neck, above the Adam's
        apple is known as the:
        A. Mandible
        B. Coronoid process
        C. Hyoid
        D. Thyroid   Ref. 3 - p. 81

385.   The names of the auditory ossicles are the:
        A. Larynx, cricoid and thyroid
        B. Navicular, cuboid and talus
        C. Canine, incisor and molar
        D. Malleus, incus and stapes   Ref. 3 - p. 81

386.   The auditory ossicles lie in what part of the body?:
        A. External ear
        B. Optic foramen
        C. Middle ear
        D. Orbital fissure   Ref. 3 - p. 81

387.   The auditory ossicles connect the _____ with the _____:
        A. Internal meatus, auditory canal
        B. Internal ear, ear drum
        C. External meatus, auditory nerve
        D. Temporal bone, sphenoid bone   Ref. 3 - p. 81

388.   How many "baby teeth" are there in the normal child?:
        A   12
        B. 20
        C. 32
        D. 40   Ref. 3 - p. 81

389.   How many permanent teeth are there in the normal adult?:
        A. 12
        B. 20
        C. 32
        D. 40   Ref. 3 - p. 81

390.   The four teeth in the front of the mouth are known as the:
        A. Molars
        B. Canines
        C. Bicuspids
        D. Incisors   Ref. 3 - p. 81

391.   The wisdom teeth are really the:
        A. Third molars
        B. Canines
        C. Incisors
        D. Premolars   Ref. 3 - p. 81

392. A tooth that grows at angles and therefore can not erupt normally is said to be:
    A. Infected
    B. Deciduous
    C. Carious
    D. Impacted   Ref. 3 - p. 82

393. The central cavity of a tooth is called the:
    A. Dental cavity
    B. Pulp cavity
    C. Incisor foramen
    D. Dental canal   Ref. 3 - p. 82

394. That part of a tooth that is imbedded in a socket of the jaw is called the:
    A. Crown
    B. Root
    C. Cusp
    D. Neck   Ref. 3 - p. 82

395. The very hard covering of the exposed part of a tooth is the:
    A. Enamel
    B. Pulp
    C. Dentine
    D. Cementum   Ref. 3 - p. 82

396. The position of opposing upper and lower teeth when the mouth is closed is known as:
    A. Cusp
    B. Function
    C. Occlusion
    D. Friction   Ref. 3 - p. 82

397. The frontal lobes of the brain lie within the:
    A. Foramen magnum
    B. Middle cranial fossa
    C. Anterior cranial fossa
    D. Sella turcica   Ref. 3 - p. 82

398. The circular opening in the back of the orbit through which passes the optic nerve is known as the:
    A. Acoustic meatus
    B. Optic foramen
    C. Foramen magnum
    D. Infraorbital foramen   Ref. 3 - p. 82

399. Which of the following foramina is not found in the base of the skull?:
    A. Foramen ovale
    B. Foramen spinosum
    C. Juglar foramen
    D. Mental foramen   Ref. 3 - p. 83

400. The flat bones of the skull, except those of the base, are preformed as:
    A. Membranes
    B. Cartilage
    C. Bone
    D. Muscle   Ref. 3 - p. 83

401. The palpable soft spots in the roof on the infant skull are called the:
A. Joints
B. Fontanelles
C. Foramina
D. Fractures   Ref. 3 - p. 83

402. A patient with an extremely small head is said to be:
A. Anencephalic
B. Hydrocephalic
C. Macrocephalic
D. Microcephalic   Ref. 3 - p. 84

403. Failure of union in the midline of the bones of the roof of the mouth results in a condition known as:
A. Cleft palate
B. Cyclops
C. Meningocele
D. Spina bifida   Ref. 3 - p. 84

404. The paired, air filled cavities in the facial and cranial bones are the:
A. Fontanelles
B. Sutures
C. Paranasal sinuses
D. Orbits   Ref. 3 - p. 84

405. The antra are also known as the _____ sinuses:
A. Maxillary
B. Sphenoid
C. Frontal
D. Ethmoid   Ref. 3 - p. 84

406. The sinuses which lie beneath the sella turcica are the:
A. Ethmoid
B. Maxillary
C. Frontal
D. Sphenoid   Ref. 3 - p. 84

407. The suture which crosses the skull transversely and separates the frontal and parietal bones is the _____ suture:
A. Lambdoid
B. Coronal
C. Sagittal
D. Squamosal   Ref. 3 - p. 84

408. The biceps tendon is attached to the:
A. Radial tuberosity
B. Lesser tubercle
C. Lateral epicondyle
D. Coronoid fossa   Ref. 3 - p. 37

409. The prominent cartilage of the ear is known as the:
A. Pinna
B. Meniscus
C. Septum
D. Canthus   Ref. 3 - p. 85

410.   The canals in the inner ear that affect the equilibrium are the:
        A. Auditory canals
        B. Semicircular canals
        C. Optic canals
        D. Eustachian tubes   Ref. 3 - p. 86

411.   An example of a hinge joint is the:
        A. Intercarpal joint
        B. Hip
        C. Shoulder
        D. Elbow   Ref. 3 - p. 92

412.   An example of a ball and socket joint is the:
        A. Intercarpal joint
        B. Metacarpo-phalangeal joint
        C. Elbow
        D. Hip   Ref. 3 - p. 93

413.   The innominate or hip bone consists of _____ bones which are
        joined at the acetabulum:
        A. Six
        B. Four
        C. Two
        D. Three   Ref. 3 - p. 46

414.   The muscle found in the wall of hollow organs, such as the stomach, is:
        A. Skeletal
        B. Involuntary
        C. Striated
        D. Voluntary   Ref. 3 - p. 94

415.   Which type of muscle can be made to function at will?:
        A. Cardiac
        B. Visceral
        C. Skeletal
        D. Non-striated   Ref. 3 - p. 94

416.   The long, cylindrical cells of skeletal muscle are called:
        A. Fibers
        B. Membranes
        C. Tendons
        D. Bursae   Ref. 3 - p. 95

417.   The connective tissue muscle sheath is known as:
        A. Tendon
        B. Fascia
        C. Cell membrane
        D. Bursa   Ref. 3 - p. 95

418.   The cord-like sinew ("leader") extending from one end of a muscle to a
        bone or ligament, is called a:
        A. Tendon
        B. Bursa
        C. Fascia
        D. Sheath   Ref. 3 - p. 95

419.   With contraction, a muscle becomes:
        A. Relaxed
        B. Longer
        C. Shorter
        D. Narrower   Ref. 3 - p. 95

420.   All of the following muscles are found in the upper extremity, except
        which one?:
        A. Biceps brachia
        B. Deltoid
        C. Triceps
        D. Sternocleidomastoid   Ref. 3 - p. 94

421.   Prolonged muscular contraction, often as a result of injury, is termed:
        A. Convulsion
        B. Paralysis
        C. Spasm
        D. Atony   Ref. 3 - p. 96

422.   Injury to, or destruction of a nerve supplying a particular muscle,
        leads to:
        A. Paralysis
        B. Spasm
        C. Sprain
        D. Convulsion   Ref. 3 - p. 96

423.   The thick, fan-shaped muscle which covers the upper, anterior chest
        wall is the:
        A. Deltoid
        B. Sternocleidomastoid
        C. Pectoralis major
        D   Diaphragm   Ref. 3 - p. 100

424.   The muscular partition between the thoracic and abdominal cavities
        is known as the:
        A. Psoas
        B. Rectus abdominis
        C. Intercostal
        D. Diaphragm   Ref. 3 - p. 100

425.   The phrenic nerve, is the nerve to the:
        A. Diaphragm
        B. Leg
        C. Shoulder
        D. Back   Ref. 3 - p. 102

426.   The fluid portion of the blood is known as:
        A. Erythrocyte
        B. Platelet
        C. Hemoglobin
        D. Plasma   Ref. 3 - p. 104

427.   Blood returned to the body tissues from the lungs will have a _____
        concentration:
        A. High carbon dioxide
        B. Low oxygen
        C. High oxygen
        D. High carbon monoxide   Ref. 3 - p. 104

428. The iron and protein compound, contained within the red blood cells, is known as:
A. Haptoglobin
B. Hemoglobin
C. Hemophilia
D. Platelet   Ref. 3 - p. 105

429. The blood cells that are specifically concerned with blood clotting are the:
A. Platelets
B. Monocytes
C. Eosinophiles
D. Basophiles   Ref. 3 - p. 106

430. A decrease in the number of red blood cells, or the amount of hemoglobin results in a condition known as:
A. Polycythemia
B. Leukocytosis
C. Anemia
D. Hyperemia   Ref. 3 - p. 106

431. _____ are thick round structures arising from the posterior surface of a vertebral body:
A. Transverse processes
B. Pedicles
C. Articular processes
D. Spinous processes   Ref. 3 - p. 60

432. The heart has how many chambers?:
A. 1
B. 2
C. 3
D. 4   Ref. 3 - p. 107

433. The special muscle of the heart is called the:
A. Pericardium
B. Myocardium
C. Endocardium
D. Myometrium   Ref. 3 - p. 107

434. The arteries that supply the heart with blood are the _____ arteries:
A. Pulmonary
B. Subclavian
C. Coronary
D. Intercostal   Ref. 3 - p. 109

435. The pulmonary artery arises from which chamber of the heart?:
A. Right atrium
B. Left atrium
C. Left ventricle
D. Right ventricle   Ref. 3 - p. 109

436. The arteries and veins which carry blood from the heart to all parts of the body and then back to the heart are called the _____ vessels:
A. Pulmonary
B. Systemic
C. Lymphatic
D. Capillary   Ref. 3 - p. 109

437.   The main artery of the systemic circulation which originates from the
       left ventricle is the:
       A.   Pulmonary conus
       B.   Pulmonary vein
       C.   Vena cava
       D.   Aorta   Ref. 3 - p. 110

438.   The tiny blood vessels which form networks between the final branches of
       arteries and veins are called:
       A.   Lymphatics
       B.   Auricles
       C.   Capillaries
       D.   Ventricles   Ref. 3 - p. 110

439.   The large trunk veins for the systemic circulation are the:
       A.   Superior and inferior vena cavae   C.   Pulmonary veins
       B.   Portal veins   D.   Hepatic veins
                                                Ref. 3 - p. 111

440.   The interarticular joints of the spinal column are known as:
       A.   Apophyseal joints
       B.   Accessory joints
       C.   Symphysis
       D.   Nucleus pulposus   Ref. 3 - p. 63

441.   The pulmonary veins drain blood from the lung to the:
       A.   Left ventricle
       B.   Right ventricle
       C.   Left atrium
       D.   Right atrium   Ref. 3 - p. 111

442.   Blood in the pulmonary artery travels from the:
       A.   Right ventricle to the lung
       B.   Lung to the left atrium
       C.   Left ventricle to the lung
       D.   Lung to the right atrium   Ref. 3 - p. 111

443.   The abdominal aorta divides into the right and left _____ artery at
       the level of the fourth lumbar vertebra:
       A.   Subclavian
       B.   Carotid
       C.   Femoral
       D.   Iliac   Ref. 3 - p. 111

444.   The system whose function, similar to the venous system, is the draining
       off of tissue fluids, is the:
       A.   Lymphatic
       B.   Arterial
       C.   Excretory
       D.   Digestive   Ref. 3 - p. 111

445.   The contracting phase of the heart cycle is known as:
       A.   Syncope
       B.   Systole
       C.   Diastole
       D.   Diastasis   Ref. 3 - p. 113

446.   When the ventricles of the heart contract which heart valves close?:
        A. Aortic
        B. Pulmonary
        C. Atrio-ventricular
        D. All of the above   Ref. 3 - p. 113

447.   The pulse represents the dilatation of which blood vessel?:
        A. Vein
        B. Venule
        C. Capillary
        D. Artery   Ref. 3 - p. 114

448.   A patient's blood pressure is 120/60; the diastolic pressure is therefore:
        A. 120
        B. 60
        C. 2
        D. 0.5   Ref. 3 - p. 114

449.   High blood pressure is known as:
        A. Systole
        B. Hypotension
        C. Hypertension
        D. Hyperbaric   Ref. 3 - p. 114

450.   The fetus obtains oxygen by way of:
        A. Its lungs
        B. The umbilical artery
        C. The foramen ovale
        D. The placenta   Ref. 3 - p. 114

451.   Inflammation of a vein is referred to as:
        A. Cellulitis
        B. Phlebitis
        C. Arteriosclerosis
        D. Varicosities   Ref. 3 - p. 115

452.   A blood clot which forms within a blood vessel is known as a(an):
        A. Thrombus
        B. Aneurysm
        C. Varicosity
        D. Aneurysm   Ref. 3 - p. 115

453.   That point on the roof of the skull where the two parietal bones meet
        the frontal bone is called:
        A. Pterion
        B. Lambda
        C   Bregma
        D. Occiput   Ref. 3 - p. 73

454.   The blood supply to the brain is via the _____ arteries:
        A. Subclavian
        B. Radial and ulnar
        C. Carotid and vertebral
        D. Iliac and femoral   Ref. 3 - p. 117

455. The celiac, superior mesenteric, and inferior mesenteric arteries are branches of the:
   A. Ascending thoracic aorta
   B. Aortic arch
   C. Descending thoracic aorta
   D. Abdominal aorta   Ref. 3 - p. 117

456. The subclavian artery supplies the:
   A. Thorax
   B. Upper extremity
   C. Lower estremity
   D. Abdomen   Ref. 3 - p. 117

457. A common site to palpate the pulse is at the:
   A. Shoulder
   B. Calf
   C. Wrist
   D. Back of the neck   Ref. 3 - p. 117

458. The popliteal artery is found in which area of the body?:
   A. Neck
   B. Upper extremity
   C. Lower extremity
   D. Abdomen   Ref. 3 - p. 117

459. The flat cartilage plate that lies below the base of the tongue and projects backwards into the pharynx is the:
   A. Uvula
   B. Epiglottis
   C. Adenoid
   D. Nares   Ref. 3 - p. 121

460. The canal connecting the nasopharynx with the middle ear is the:
   A. Eustachian tube
   B. Nasal cavity
   C. Epiglottis
   D. Auditory canal   Ref. 3 - p. 121

461. The cartilage plate lying in the front of the neck, commonly referred to as the "adam's apple, " is the:
   A. Trachea
   B. Cricoid cartilage
   C. Thyroid cartilage
   D. Epiglottis   Ref. 3 - p. 121

462. The hollow, tube-like structure connecting the larynx and main stem bronchi is the:
   A. Nasopharynx
   B. Oropharynx
   C. Aorta
   D. Trachea   Ref. 3 - p. 121

463. The following is not a facial bone:
   A. Temporal
   B. Maxilla
   C. Nasal
   D. Zygomatic   Ref. 3 - p. 73

464. The right and left lungs are separated by the:
     A. Diaphragm
     B. Thoracic spine
     C. Mediastinum
     D. Sternum   Ref. 3 - p. 122

465. The tapered upper end of the lung is called the:
     A. Apex
     B. Base
     C. Lobe
     D. Fissure   Ref. 3 - p. 122

466. How many lobes make up the right lung?:
     A. 2
     B. 3
     C. 4
     D. 5   Ref. 3 - p. 122

467. How many lobes make up the left lung?:
     A. 2
     B. 3
     C. 4
     D. 5   Ref. 3 - p. 122

468. The depression on the medial surface of the lung where vessels and bronchi enter and leave the lung is known as the:
     A. Fissure
     B. Hilum
     C. Lobule
     D. Pleura   Ref. 3 - p. 122

469. The covering membrane of the lung is the:
     A. Pericardium
     B. Peritoneum
     C. Pleura
     D. Dura   Ref. 3 - p. 123

470. The triangular space at the junction of the diaphragm and lateral chest wall is the:
     A. Costophrenic sulcus
     B. Subglottic space
     C. Apex
     D. Lung root   Ref. 3 - p. 123

471. During inspiration, the diaphragm is:
     A. Inverted
     B. Displaced downward
     C. Elevated
     D. Obliterated   Ref. 3 - p. 123

472. The usual rate of respiration is about _____ times per minute:
     A. Three
     B. Eighteen
     C. Thirty
     D. Sixty   Ref. 3 - p. 123

473.   The _____ bone forms the central inferior border of the middle
       cranial fossa:
       A. Zygomatic
       B. Temporal
       C. Sphenoid
       D. Occipital   Ref. 3 - p. 82

474.   Abnormal dilatation of the bronchi is known as:
       A. Bronchitis
       B. Bronchiolitis
       C. Bronchiectasis
       D. Bronchopneumonia   Ref. 3 - p. 123

475.   Free air in the pleural cavity is known as:
       A. Empyema
       B. Hemothorax
       C. Pleurisy
       D. Pneumothorax   Ref. 3 - p. 123

476.   The lining membrane of the abdominal and pelvic cavities is the:
       A. Pleura
       B. Pericardium
       C. Peritoneum
       D. Perineum   Ref. 3 - p. 126

477.   A thick ring of smooth muscle around the opening of a hollow organ,
       such as the stomach or anus, is known as a:
       A. Sphincter
       B. Peristalsis
       C. Hernia
       D. Lumen   Ref. 3 - p. 126

478.   Waves of alternating dilatation and contraction in the wall of a hollow
       organ such as the stomach or intestine, are referred to as _____
       waves:
       A. Peritoneal
       B. Peristaltic
       C. Spastic
       D. Atonic   Ref. 3 - p. 127

479.   The parotid, submandibular and sublingual glands are known collectively
       as the:
       A. Ductless glands
       B. Endocrine glands
       C. Lymph glands
       D. Salivary glands   Ref. 3 - p. 127

480.   The parotid glands are located:
       A. Behind the ear
       B. In front of and below the ear
       C. In the floor of the mouth
       D. In the pharynx   Ref. 3 - p. 127

481.   Small grooves on the inner surface of the skull radiating up from the
       foramen spinosum accommodate branches of the:
       A. Middle cerebral artery
       B. Middle meningeal artery
       C. Anterior cerebral artery
       D. Posterior cerebral artery   Ref. 3 - p. 83

482. The opening in the upper end of the stomach is the:
   A. Cardiac orifice
   B. Pylorus
   C. Incisura
   D. Colostomy   Ref. 3 - p. 128

483. The opening at the lower end of the stomach is the:
   A. Cardiac orifice
   B. Pylorus
   C. Incisura
   D. Colostomy   Ref. 3 - p. 128

484. The stomach is located in which quadrant of the abdomen?:
   A. Right upper
   B. Left upper
   C. Right lower
   D. Left lower   Ref. 3 - p. 128

485. The large, convex left border of the stomach is called the:
   A. Pylorus
   B. Lesser curvature
   C. Greater curvature
   D. Fundus   Ref. 3 - p. 128

486. The first portion of the small intestine is known as the:
   A. Pylorus
   B. Duodenum
   C. Jejunum
   D. Ileum   Ref. 3 - p. 128

487. The organ which lies within the curve of the duodenal loop is the:
   A. Spleen
   B. Liver
   C. Gall bladder
   D. Pancreas   Ref. 3 - p. 128

488. The second portion of the small intestine which lies in the upper and mid-abdomen is the:
   A. Pylorus
   B. Fundus
   C. Ileum
   D. Jejunum   Ref. 3 - p. 130

489. The third portion of the small intestine, that joins with the colon, is the:
   A. Pylorus
   B. Duodenum
   C. Jejunum
   D. Ileum   Ref. 3 - p. 130

490. The junction of the large and small intestine is at the:
   A. Ileo-cecal valve
   B. Pylorus
   C. Esophago-gastric junction
   D. Hepatic flexure   Ref. 3 - p. 130

491. The pouch-like, blind end portion of the colon is known as the:
   A. Sigmoid
   B. Duodenum
   C. Fundus
   D. Cecum   Ref. 3 - p. 130

492. The bend in the colon between the ascending and transverse portions is the:
    A. Sigmoid
    B. Hepatic flexure
    C. Splenic flexure
    D. Anus   Ref. 3 - p. 130

493. The bend in the colon at the junction of the transverse and descending portion is the:
    A. Sigmoid
    B. Hepatic flexure
    C. Splenic flexure
    D. Anus   Ref. 3 - p. 130

494. The "S"-shaped portion of the descending colon is called the:
    A. Hepatic flexure
    B. Splenic flexure
    C. Cecum
    D. Sigmoid   Ref. 3 - p. 130

495. The final portion of the colon is called the:
    A. Rectum
    B. Cecum
    C. Sigmoid
    D. Buttock   Ref. 3 - p. 130

496. In order to properly insert an enema tip, it is important to remember that the anal canal extends _____ for a distance of about 1-1/2 inches:
    A. Up and backward
    B. Up and forward
    C. Up and to the right
    D. Up and to the left   Ref. 3 - p. 130

497. The pancreas lies:
    A. Behind the stomach and duodenum
    B. In the lower abdomen
    C. Above the diaphragm
    D. In the pelvis   Ref. 3 - p. 130

498. Insulin is secreted by which organ?:
    A. Gall bladder
    B. Pancreas
    C. Liver
    D. Stomach   Ref. 3 - p. 132

499. A lack of insulin in the body results in what condition?:
    A. Diabetes
    B. Hepatitis
    C. Cushing's disease
    D. Ulcerative colitis   Ref. 3 - p. 132

500. The largest solid organ in the body is the:
    A. Spleen
    B. Pancreas
    C. Colon
    D. Liver   Ref. 3 - p. 132

501. The gall bladder lies in:
    A. The left upper quadrant
    B. The right lower quadrant
    C. An impression on the undersurface of the liver
    D. The duodenal loop   Ref. 3 - p. 132

502. The duct of the gall bladder is called the:
A. Hepatic duct
B. Cystic duct
C. Common duct
D. Pancreatic duct   Ref. 3 - p. 132

503. The hepatic and cystic ducts unite to form the:
A. Pancreatic duct
B. Accessory pancreatic duct
C. Common duct
D. Gall bladder   Ref. 3 - p. 132

504. The common bile duct opens into the duodenum at the:
A. Ampulla of Vater
B. Gall bladder
C. Ligament of Treitz
D. Omentum   Ref. 3 - p. 132

505. A congenital diverticulum of the ileum, located about 3 feet above the ileo-cecal junction, is known as a _____ diverticulum:
A. Zenker's
B. Meckel's
C. Duodenal
D. Jejunal   Ref. 3 - p. 133

506. Stones in the gall bladder are known as:
A. Nephrolithiasis
B. Diverticula
C. Cholelithiasis
D. Cholecystogram   Ref. 3 - p. 134

507. Outpouchings through the wall of a hollow organ, such as the colon, are called:
A. Fistulae
B. Diverticula
C. Hepatitis
D. Hernia   Ref. 3 - p. 134

508. At rest a muscle remains partially contracted. This is called:
A. Muscle tone
B. Flexion
C. Spasm
D. Atony   Ref. 3 - p. 95

509. Surgical removal of the gall bladder is called:
A. Cholelithiasis
B. Cholecystitis
C. Cholecystectomy
D. Cholecystogram   Ref. 3 - p. 134

510. Partial or total surgical removal of the stomach is called a:
A. Gastrectomy
B. Gastritis
C. Gastrectasia
D. Gastroenterostomy   Ref. 3 - p. 134

511.   The tablets given to a patient for a gall bladder examination are
        excreted by the:
        A. Pancreas
        B. Adrenal
        C. Spleen
        D. Liver   Ref. 3 - p. 135

512.   One function of the gall bladder is the absorption of:
        A. Protein
        B. Hydrochloric acid
        C. Oxygen
        D. Water   Ref. 3 - p. 136

513.   The presence of fat in the duodenum causes the gall bladder to:
        A. Fill with bile
        B. Excrete iodine
        C. Contract
        D. Dilate   Ref. 3 - p. 136

514.   Which of the following factors may result in nonvisualization of the gall
        bladder on an oral study?:
        A. Liver disease
        B. Gall bladder disease
        C. Obstruction of the stomach
        D. All of the above   Ref. 3 - p. 136

515.   The kidneys occupy the space between which of the following levels?:
        A. C5-C7
        B. D7-D10
        C. D12-L3
        D. L4-S1   Ref. 3 - p. 139

516.   The ureters extend from the kidneys to the:
        A. Urinary bladder
        B. Adrenal gland
        C. Urethra
        D. Gall bladder   Ref. 3 - p. 139

517.   The upper, expanded, funnel-shaped end of the ureter which is in
        contact with the kidney is known as the:
        A. Bladder
        B. Urethra
        C. Calyx
        D. Renal pelvis   Ref. 3 - p. 139

518.   The urinary bladder lies:
        A. Behind the rectum
        B. In front of the rectum
        C. In front of the sacrum
        D. In back of the sacrum   Ref. 3 - p. 139

519.   The urethra is the passage from the urinary bladder to the:
        A. Kidney
        B. Renal pelvis
        C. Outside
        D. Abdomen   Ref. 3 - p. 139

520. In the male, the urethra is _____ the urethra in the female:
A. Longer than
B. Shorter than
C. The same size as   Ref. 3 - p. 139

521. The blood supply to the kidneys is via the:
A. Iliac arteries
B. Renal arteries
C. Renal tubules
D. Prostatic plexus   Ref. 3 - p. 139

522. The outer part of the kidney is known as the:
A. Pelvis
B. Tubule
C. Medulla
D Cortex   Ref. 3 - p. 139

523. The muscle passing from the mastoid process to the medial end of the clavicle and upper sternum is the _____ muscle:
A. Trapezius
B. Pectoralis major
C. Sternocleidomastoid
D. Omohyoid   Ref. 3 - p. 100

524. The accumulation of harmful waste products in the body because of malfunction of the kidneys is referred to as:
A. Uremia
B. Urination
C. Cystitis
D. Polyuria   Ref. 3 - p. 140

525. The kidneys lie in which portion of the body?:
A. Pelvis
B. Thorax
C. Posterior abdomen
D. Anterior abdomen   Ref. 3 - p. 141

526. The female reproductive organ is the:
A. Ureter
B. Ovary
C. Prostate
D. Oviduct   Ref. 3 - p. 142

527. The hollow tubes which pass from the upper outer margin of the uterus are the:
A. Ureters
B. Seminal vesicles
C. Fallopian tubes
D. Vagina   Ref. 3 - p. 142

528. In the female, the fertilized ovum develops in the:
A. Uterus
B. Vagina
C. Peritoneum
D. Ovary   Ref. 3 - p. 142

529. The lower, constricted part of the uterus is known as the:
A. Vagina
B. Fundus
C. Cervix
D. Ovary　　　　　　　　　　Ref. 3 - p. 143

530. The muscle which may be demonstrated on abdominal X-ray as an oblique line running from the first lumbar vertebra to the iliac crest is the:
A. Psoas muscle
B. Sacrospinalis muscle
C. Brachialis muscle
D. Rectus abdominis muscle　　　Ref. 3 - p. 102

531. Cancer of the breast may spread through the lymphatic vessels to the:
A. Arm
B. Axilla
C. Uterus
D. Diaphragm　　　　　　　Ref. 3 - p. 144

532. The male reproductive organs are called the:
A. Prostate
B. Ovaries
C. Spermatozoa
D. Testes　　　　　　　　Ref. 3 - p. 146

533. The globular gland lying below the outlet of the urinary bladder in the male is the:
A. Prostate
B. Testis
C. Ovary
D. Scrotum　　　　　　　Ref. 3 - p. 147

534. Ductless glands are also known as the _____ glands:
A. Salivary
B. Digestive
C. Endocrine
D. Reproductive　　　　　　Ref. 3 - p. 149

535. A secretion manufactured in a ductless gland is called a(an):
A. Enzyme
B. Hormone
C. Colloid
D. Suspension　　　　　　Ref. 3 - p. 149

536. Chemical changes occurring in body cells as a result of cellular activity are part of the process of:
A. Degradation
B. Fertilization
C. Secretion
D. Metabolism　　　　　　Ref. 3 - p. 149

537. The pituitary gland lies in the:
A. Sella turcica
B. Mediastinum
C. Renal fossa
D. Sphenoid sinus　　　　　Ref. 3 - p. 149

538.   The small gland which lies in the midline of the brain, is often calcified
        and therefore visible on plain X-rays of the skull, is called the:
        A. Pituitary
        B. Pineal
        C. Parathyroid
        D. Medulla   Ref. 3 - p. 150

539.   The bilobed gland which lies in the front of the neck is the:
        A. Thymus
        B. Pineal
        C. Thyroid
        D. Adrenal   Ref. 3 - p. 150

540.   Basal metabolism tests indicate the function of which gland?:
        A. Thymus
        B. Thyroid
        C. Adrenal
        D. Parathyroid   Ref. 3 - p. 151

541.   The gland which is quite large in infants and young children and eventually
        becomes much smaller with age, is the:
        A. Thymus
        B. Thyroid
        C. Parathyroid
        D. Adrenal   Ref. 3 - p. 151

542.   The glands which lie above each kidney are the:
        A. Parathyroids
        B. Ovaries
        C. Testes
        D. Adrenals   Ref. 3 - p. 151

543.   A condition resulting from overstimulation of the pituitary gland, and
        characterized by enlargement of the skull, hands and feet, is called:
        A. Cretinism
        B. Acromegaly
        C. Pneumonia
        D. Hepatitis   Ref. 3 - p. 152

544.   The term "goiter" refers to enlargement of which gland?:
        A. Parathyroid
        B. Thymus
        C. Thyroid
        D. Pituitary   Ref. 3 - p. 152

545.   The longitudinal fissure divides the brain into two halves called the:
        A. Medulla
        B. Mid-brain
        C. Cerebellum
        D. Cerebral hemispheres   Ref. 3 - p. 155

546.   The outer surface of the brain consists of numerous ridges or folds
        called:
        A. Convolutions
        B. Fissures
        C. Ventricles
        D. Sulci   Ref. 3 - p. 155

547.   Each half of the brain is divided into five:
        A. Alveoli
        B. Lobes
        C. Nerves
        D. Phalanges   Ref. 3 - p. 156

548.   The nerve centers for sensations such as speech, sight and hearing are
        contained in which division of the brain?:
        A. Cerebellum
        B. Midbrain
        C. Medulla
        D. Cerebrum   Ref. 3 - p. 156

549.   The nerve centers in the brain controlling muscular movement are
        known as:
        A. Motor
        B. Sensory
        C. Reflex
        D. Ganglion   Ref. 3 - p. 156

550.   That portion of the brain occupying the posterior cranial fossa is the:
        A. Frontal lobe
        B. Temporal lobe
        C. Cerebrum
        D. Cerebellum   Ref. 3 - p. 156

551.   The spinal cord extends from the level of the foramen magnum to the:
        A. 5th cervical vertebra
        B. 1st dorsal vertebra
        C. 12th dorsal vertebra
        D. 2nd lumbar vertebra   Ref. 3 - p. 156

552.   The spinal nerves extending down the spinal canal below the mid-lumbar
        level are referred to as the:
        A. Medulla
        B. Pons
        C. Cauda equina
        D. Ganglia   Ref. 3 - p. 156

553.   Each spinal nerve leaves the spinal canal through the:
        A. Foramen magnum
        B. Intervertebral foramen
        C. Central canal
        D. Subarachnoid space   Ref. 3 - p. 156

554.   There are how many pairs of cranial nerves?:
        A. 4
        B. 12
        C. 24
        D. 32   Ref. 3 - p. 158

555.   The cranial nerves exit via the:
        A. Intervertebral foramina
        B. Foramina in the skull
        C. Central canal
        D. Spinal canal   Ref. 3 - p. 158

556.   The coverings of the brain and spinal cord are the:
    A.   Pleura
    B.   Meninges
    C.   Ventricles
    D.   Vagus   Ref. 3 - p. 158

557.   The _____ valve prevents back flow of blood from the aorta to the ventricle:
    A.   Aortic
    B.   Mitral
    C.   Tricuspid
    D.   Pulmonary   Ref. 3 - p. 109

558.   The blood vessel that returns oxygenated blood from the placenta to the fetus:
    A.   Umbilical artery
    B.   Umbilical vein
    C.   Ductus arteriosus
    D.   Pulmonary vein   Ref. 3 - p. 114

559.   _____ are collections of lymphoid tissue on the roof and posterior wall of the nasopharynx:
    A.   Glands
    B.   Nodes
    C.   Tonsils
    D.   Adenoids   Ref. 3 - p. 121

560.   The potential space between the lung and chest wall is the:
    A.   Pleural cavity
    B.   Pulmonary cavity
    C.   Pneumonic cavity
    D.   Dural cavity   Ref. 3 - p. 123

561.   The double layer of peritoneum which supports the abdominal organs is called:
    A.   Membrane
    B.   Omentum
    C.   Mesentery
    D.   Ligament   Ref. 3 - p. 126

562.   The function of the kidneys is to:
    A.   Excrete waste products
    B.   Regulate fluid content of blood
    C.   Regulate salt content of blood
    D.   All of the above   Ref. 3 - p. 140

563.   A pregnancy occurring outside of the uterus is known as:
    A.   Embryonic
    B.   Heterotopic
    C.   Ectopic
    D.   Atrophic   Ref. 3 - p. 144

564.   The _____ glands regulate the calcium content of the blood:
    A.   Parathyroid
    B.   Thyroid
    C.   Adrenal
    D.   Pituitary   Ref. 3 - p. 150

565.   Myxedema is a disease due to decreased function or removal of the
           _____ gland:
       A.   Pineal
       B.   Adrenal
       C.   Thyroid
       D.   Parathyroid   Ref. 3 - p. 152

566.   What gland is often seen in chest X-rays of infants?:
       A.   Thymus
       B.   Thyroid
       C.   Adrenal
       D.   Parathyroid   Ref. 3 - p. 153

567.   The structure through which fibers pass from one cerebral hemisphere
       to the other is the:
       A.   Medulla oblongata
       B.   Transverse fissure
       C.   Falx cerebri
       D.   Corpus callosum   Ref. 3 - p. 155

568.   The trigeminal nerve is a sensory cranial nerve to the _____ :
       A.   Face
       B.   Neck
       C.   Eye muscles
       D.   Pharynx   Ref. 3 - p. 158

569.   The openings in the fourth ventricle which allow cerebral spinal fluid to
       pass from the ventricle into the subarachnoid space are called the
       foramina of _____ :
       A.   Rotundum
       B.   Spinosum
       C.   Monroe
       D.   Luschka and Magendie   Ref. 3 - p. 159

570.   Inflammation of the brain is called:
       A.   Neuritis
       B.   Encephalitis
       C.   Meningitis
       D.   Myelitis   Ref. 3 - p. 160

FOR EACH OF THE FOLLOWING MULTIPLE CHOICE QUESTIONS
SELECT THE ONE MOST APPROPRIATE ANSWER:

571.   Synarthrodial joints are found in the:
       A. Hip
       B. Shoulder
       C. Skull
       D. Spine   Ref. 4 - p. 24

572.   Freely movable joints are termed:
       A. Synarthrodial
       B. Amphiarthrodial
       C. Diarthrodial   Ref. 4 - p. 24

573.   The head of the radius articulates with the _____ of the humerus:
       A. Capitellium
       B. Trachlea
       C. Epicondyle
       D. Semilunar notch   Ref. 4 - p. 30

574.   In the lateral view of the second (index) finger the hand should rest on
       its _____ surface:
       A. Ulnar
       B. Radial
       C. Dorsal   Ref. 4 - p. 37

575.   Which of the views of the thumb gives the best delineation of the first
       carpometacarpal joint?:
       A. Posterior anterior
       B. Anterioposterior
       C. Lateral   Ref. 4 - p. 38

576.   The "carpal bridge" view is recommended for the demonstration of:
       A. Lunate dislocations
       B. Foreign bodies in the wrist
       C. Navicular fractures
       D. All of the above   Ref. 4 - p. 46

577.   The frontal view obtained with the elbow in acute flexion gives a clear
       view of the:
       A. Epicondyles
       B. Capitellium
       C. Trochlea
       D. Olecranon process   Ref. 4 - p. 51

578.   The accuracy of the lateral view of the arm or elbow shown by super-
       imposition of the:
       A. Epicondyles of the humerus
       B. Tuberosities of the humerus
       C. Olecranon and radial head   Ref. 4 - p. 56

579.   The skull, vertebrae, sternum and ribs are collectively referred to as the:
       A. Regional skeleton
       B. Axial skeleton
       C. Appendicular skeleton
       D. Long bones   Ref. 4 - p. 22

580. The extremities, their girdles and the pelvis are collectively referred
    to as:
    A. Regional skeleton
    B. Axial skeleton
    C. Appendicular skeleton
    D. Long bones   Ref. 4 - p. 22

581. The supine position refers to the patient lying in which position?:
    A. On the back
    B. Face down
    C. Right side down
    D. With the feet elevated   Ref. 4 - p. 25

582. The prone position refers to the patient lying in which position?:
    A. On the back
    B. Face down
    C. Right side down
    D. Feet elevated   Ref. 4 - p. 25

583. The lateral recumbent position refers to the patient lying:
    A. Face down
    B. On the back
    C. On the side
    D. Oblique   Ref. 4 - p. 25

584. The term extension refers to _____ of a joint:
    A. Bending
    B. Turning inward
    C. Straightening
    D. Turning outward   Ref. 4 - p. 25

585. In abduction, the part is moved _____ the central axis of the body:
    A. Upward from
    B. Downward from
    C. Away from
    D. Toward   Ref. 4 - p. 25

586. In adduction, the part is moved _____ the central axis of the body:
    A. Upward from
    B. Downward from
    C. Away from
    D. Toward   Ref. 4 - p. 25

587. Posterior, or dorsal, designates the _____ part of the body or
    organ:
    A. Front
    B. Back
    C. Side
    D. Top   Ref. 4 - p. 25

588. Caudal refers to a part that is _____ the head of the body:
    A. Away from
    B. Toward
    C. In back of
    D. In front of   Ref. 4 - p. 25

589. Distal refers to a part that is _____ the beginning or origin of a
      structure:
      A. Adjacent to
      B. On top of
      C. Away from
      D. Near   Ref. 4 - p. 25

590. Proximal refers to a part that is _____ the beginning or origin of
      structure:
      A. Adjacent to
      B. On top of
      C. Away from
      D. Near   Ref. 4 - p. 25

591. The fluid filled sacs found in relation to moving joints are called:
      A. Ligaments
      B. Bursae
      C. Tendons
      D. Menisci   Ref. 4 - p. 24

592. When screens are utilized in radiography of the foot it is advisable to use:
      A. High MaS - low kilovoltage
      B. High kilovoltage - low MaS
      C. High kilovoltage - high MaS   Ref. 4 - p. 60

593. To demonstrate the sesamoid bones of the first metatarsal a(an)
      _____ view is used:
      A. Axial
      B. Oblique
      C. Frontal   Ref. 4 - p. 68

594. In order to demonstrate any forward or backword displacement in
      fractures of the metacarpal bones, the following view is useful:
      A. A P view of the hand
      B. Lateral view of the hand
      C. A P view of wrist
      D. Lateral view of forearm   Ref. 4 - p. 33

595. A lateral view of the hand is usually obtained with the part in normal:
      A. Flexion
      B. Extension
      C. Supination
      D. Pronation   Ref. 4 - p. 33

596. To identify opaque foreign bodies in the hand, the following view is
      customary:
      A. Overexposed A P
      B. Chassard-Lapine
      C. Lateral in extension
      D. Tunnel view   Ref. 4 - p. 34

597. The oblique projection of the hand is obtained using a _____ obliquity:
      A. 35 degrees
      B. 45 degrees
      C. 60 degrees
      D. 75 degrees   Ref. 4 - p. 35

598.   The inter-carpal spaces are better demonstrated using which projection?:
        A. A P wrist
        B. P A wrist
        C. Lateral wrist
        D. Oblique wrist   Ref. 4 - p. 40

599.   To better view the carpal navicular bone in the A P view, which maneuver
        is used?:
        A. Flexing the fingers
        B. Radial deviation (ulner flexion)
        C. Extension of wrist
        D. Flexion of wrist   Ref. 4 - p. 42

600.   The view used to separate the pisiform bone from the remaining carpal
        bones is:
        A. P A wrist
        B. A P wrist
        C. Anterior oblique of wrist
        D. Lateral wrist   Ref. 4 - p. 43

601.   In the Stecher position to visualize the carpal navicular, the wrist is
        angled _____ degrees to the horizontal plane:
        A. 10
        B. 20
        C. 30
        D. 40   Ref. 4 - p. 44

602.   The same view can also be obtained if the wrist is kept flat and the tube
        is angled toward the:
        A. Fingers
        B. Elbow
        C. Ulna
        D. Radius   Ref. 4 - p. 44

603.   In the carpal tunnel view the wrist is hyperextended and the central ray
        angled _____ degrees toward the palm of the hand:
        A. 20
        B. 30
        C. 40
        D. 50   Ref. 4 - p. 45

604.   To determine the position of fragments of metatarsal fractures a(an)
        _____ view is obtained:
        A. Axial
        B. Lateral
        C. Anterior posterior
        D. Oblique   Ref. 4 - p. 70

605.   In the right anterior oblique projection of the chest, the sternum is
        projected to the _____ of the vertebrae, so as to overshadow the
        _____ hemithorax:
        A. Left, left
        B. Right, right
        C. Left, right
        D. Right, left   Ref. 7 - p. 236

606. When examining the lowest 4 pairs of ribs, it is best to project them:
A. Below the level of the diaphragm
B. Above the level of the diaphragm
C. Laterally
D. In flexion-extension   Ref. 4 - p. 246

607. The jugular foramen is best demonstrated using which view?:
A. Stereo lateral of skull
B. Lateral of sacrum
C. Modified 20 degrees submento-vertex projection of skull
D. Oblique of cervical spine   Ref. 7 - p. 270

608. Examination of the temporo-mandibular joints normally consists of:
A. Single closed mouth view
B. Film with jaws both opened and closed
C. Single closed mouth view
D. Films while patient chews gum   Ref. 4 - p. 302

609. The Trendelenburg position consists of the patient:
A. Supine with feet elevated
B. Supine with head elevated
C. In knee-chest position
D. Lying on the side   Ref. 4 - p. 286

610. In the A P projection of the forearm or elbow, the proximal radius and ulna are partially obscured by overlap unless the film is taken with the:
A. Hand pronated
B. Hand supinated
C. Wrist flexed
D. Elbow flexed   Ref. 4 - pp. 47, 48

611. In a 45 degree oblique projection of the tarsus the central ray is _____ and directed to a point just distal to the _____:
A. Vertical, medial malleolus
B. Oblique 45°, lateral malleolus
C. Horizontal, medial malleolus
D. Vertical, lateral malleolus   Ref. 4 - p. 82

612. In a patient with an injured shoulder who cannot abduct the affected arm, the upper humerus and shoulder can still be visualized by using the:
A. Axial projection
B. Transthoracic lateral projection
C. Internal rotation view
D. Oblique projection   Ref. 4 - p. 58

613. There are _____ tarsal bones in the ankle:
A. 3
B. 5
C. 7
D. 9   Ref. 4 - p. 61

614. The largest tarsal bone, also known as the "heel bone," is the:
A. Cuneiform
B. Cuboid
C. Talus
D. Calcaneus   Ref. 4 - p. 61

615.   The cuboid bone lies on the _____ side of the foot:
       A.   Lateral
       B.   Medial
       C.   Inferior
       D.   Superior   Ref. 4 - p. 61

616.   Two small bones always found lying adjacent to the plantar surface of the
       first metatarso-phalangeal joint, are the:
       A.   Epiphyses
       B.   Sesamoids
       C.   Phalanges
       D.   Cuneiforms   Ref. 4 - p. 61

617.   The ankle joint is formed by the articulation of the:
       A.   Talus with the calcaneus
       B.   Calcaneus with the cuboid
       C.   Talus with the tibia and fibula
       D.   Talus with the navicular   Ref. 4 - p. 62

618.   The lateral malleolus is part of the:
       A.   Fibula
       B.   Tibia
       C.   Talus
       D.   Calcaneus   Ref. 4 - p. 62

619.   The two large protuberances on the distal end of the femur are called the:
       A.   Tuberosities
       B.   Trochanters
       C.   Condyles
       D.   Styloids   Ref. 4 - p. 63

620.   The largest sesamoid bone in the body is the:
       A.   Femur
       B.   Pelvis
       C.   Calcaneus
       D.   Patella   Ref. 4 - p. 63

621.   In the oblique view of the toes and forefoot, the central ray and ball of
       the foot form an angle of _____ degrees:
       A.   15
       B.   25
       C.   40
       D.   60   Ref. 4 - p. 66

622.   The Broden positions are utilized to demonstrate:
       A.   Posterior subtalar joint
       B.   Anterior subtalar joint
       C.   Calcaneocuboid joint
       D.   Talonavicular joint   Ref. 4 - pp. 84-85

623.   To best examine the structural status of the plantar arch, which view
       is used?:
       A.   Lateral of hand
       B.   Open mouth
       C.   Lateral weight bearing of foot
       D.   Lateral cervical spine   Ref. 4 - p. 76

624. With the tube tilted caudally, so that the central ray forms an angle of
45 degrees to the long axis of the sole of the foot, one obtains a(an):
A. Axial view of the os calcis
B. Oblique view of the tarsal bones
C. Sesamoid view
D. Occlusal view   Ref. 4 - p. 81

625. To best visualize the talo-fibular articulation, the foot is dorsiflexed and
rotated:
A. 45 degrees laterally
B. 45 degrees medially
C. 10 degrees laterally
D. 10 degrees medially   Ref. 4 - p. 90

626. With the central ray perpendicular to the film, and the leg and foot
rotated medially 45 degrees, one obtains a(an):
A. Lateral view of ankle
B. Dorsa-plantar view
C. Axial view of ankle
D. Oblique view of ankle   Ref. 4 - p. 90

627. In the A P view of the knee, to better visualize the joint space, the
joint space, the central ray may be angled 5-10 degrees:
A. Laterally
B. Medially
C. Toward the head
D. Toward the feet   Ref. 4 - p. 94

628. In the lateral view of the knee, the joint space will not be obscured by
the medial femoral condyle if the central ray is angled slightly:
A. Cephalad
B. Caudad
C. Medially
D. Laterally   Ref. 4 - p. 95

629. To visualize loose bodies in the knee joint space, one must obtain a
profile view of the:
A. Femur
B. Tibia
C. Patella
D. Intercondyloid fossa   Ref. 4 - p. 99

630. With the patient prone and the affected knee fully flexed with the patella
at right angles to the film, one obtains a(an):
A. Axial view of the patella
B. Intercondyloid view
C. P A view of patella
D. Supero-inferior view of patella   Ref. 4 - p. 104

631. The humeral head articulates with the _____ of the scapula:
A. Acetabular fossa
B. Glenoid fossa
C. Rhomboid fossa
D. Supraspinatous fossa   Ref. 4 - p. 122

632. The two prominent processes on the upper end of the humerus are the:
A. Greater and lesser tuberosities
B. Greater and lesser trochanters
C. Medial and lateral malleolus
D. Anterior and posterior tubercle   Ref. 4 - p. 122

633.   Two separate A P views of the shoulder are often obtained. They are:
        A.   Flexion and extension
        B.   Weight bearing and non-weight bearing
        C.   Abduction and adduction
        D.   Internal and external rotation   Ref. 4 - p. 124

634.   With the patient supine, the arm abducted and the central ray directed
        horizontally through the axilla, one obtains a(an):
        A.   Infero-superior projection of the shoulder
        B.   Infraspinatous insertion view of shoulder
        C.   Axial view of the shoulder joint
        D.   Bicipital groove view   Ref. 4 - p. 133

635.   To best evaluate acromioclavicular joint separation or dislocation,
        an erect PA projection is obtained with the patient's:
        A.   Holding a sandbag in each hand
        B.   Hands on hips
        C.   Neck extended
        D.   Hands over his head   Ref. 4 - p. 136

636.   If the clavicle is examined for a fracture or destructive disease, the
        following projection should not be used:
        A.   Supine
        B.   Prone
        C.   Erect
        D.   Axial   Ref. 4 - p. 138

637.   In the AP view of the ankle slight _____ of the foot will "open up"
        the talofibular articulation:
        A.   Plantar flexion
        B.   Eversion
        C.   Inversion
        D.   Pronation   Ref. 4 - p. 88

638.   For the AP projection for the scapula, the patient is supine, the central
        ray is vertical to the film and the patient's arm is:
        A.   Flexed
        B.   Internally rotated
        C.   Adducted
        D.   Abducted   Ref. 4 - p. 144

639.   To demonstrate the spine of the scapula the patient is placed supine,
        the opposite shoulder is elevated and the central ray is centered through
        the postero-superior portion of the scapula and angled:
        A.   40-45 degrees toward the head
        B.   40-45 degrees toward the feet
        C.   5-10 degrees toward the head
        D.   5-10 degrees toward the feet   Ref. 4 - p. 148

640.   To project the scapula free of bony superimposition, the following
        projection may be used:
        A.   Posterior oblique
        B.   Antero-posterior
        C.   Postero-anterior
        D.   Semi-axial AP   Ref. 4 - p. 146

641.   The bony thorax is formed by the sternum, the ribs and the:
        A.   Lumbar vertebrae
        B.   Dorsal vertebrae
        C.   Cervical vertebrae
        D.   All of the above   Ref. 4 - p. 152

642. The sternum consists of 3 parts: the body, the manubrium and the:
A. Gladiolus
B. Pedicle
C. Xiphoid
D. Lamina   Ref. 4 - p. 152

643. The body of the sternum and the manubrium are joined and form the:
A. Xiphoid
B. Sternal angle
C. Clavicle
D. Manubrial notch   Ref. 4 - p. 153

644. Because of the overlap with the dorsal vertebrae, it is impossible to obtain the following view of the sternum without using laminography:
A. Lateral
B. Latero-medial
C. Oblique
D. Direct frontal   Ref. 4 - p. 154

645. In radiographing the sternum, the overlying rib shadows can be blurred by using:
A. Short focal film distance
B. Long focal film distance
C. Overpenetrated Bucky technique
D. Barium swallow   Ref. 4 - p. 154

646. In radiographing the sternum, the shadows of the overlying pulmonary markings can be blurred out by having the patient:
A. Suspend respiration
B. Breath shallowly
C. Move during exposure
D. Lie flat on the table   Ref. 4 - p. 154

647. The oblique projection of the sternum is obtained with the patient in the _____ position:
A. Supine
B. Decubitus
C. Prone
D. Sitting   Ref. 4 - p. 155

648. A more accurate latero-medial projection of the sternum is obtained using:
A. Tube tilt method
B. Body rotation method
C. Postero-anterior projection
D. Bucky technique   Ref. 4 - p. 156

649. For the lateral erect projection of the sternum, the patient's shoulders are:
A. Rotated forward
B. Elevated
C. Depressed
D. Rotated backward   Ref. 4 - p. 158

650. For the oblique view of the sternoclavicular articulations, the patient is seated erect or placed recumbent with the body obliqued approximately:
A. 15 degrees
B. 45 degrees
C. 20 degrees
D. 50 degrees   Ref. 4 - p. 161

651.   The anterior portions of the ribs should be examined in the _____
        position:
        A.   Postero-anterior
        B.   Antero-posterior
        C.   Lateral
        D.   Apical lordotic   Ref. 4 - p. 166

652.   The axillary portions of the ribs are best examined in the _____
        position:
        A.   Lateral
        B.   Oblique
        C.   Apical lordotic
        D.   Lateral decubitus   Ref. 4 - p. 166

653.   The left rib cage may be cleared of the heart shadow by using the _____
        position:
        A.   Right anterior oblique
        B.   Left posterior oblique
        C.   Left anterior oblique
        D.   Antero-posterior   Ref. 4 - p. 167

654.   Better detail will be obtained when examining the ribs above the diaphragm
        if the following maneuver is used:
        A.   Shallow breathing
        B.   Suspended respiration
        C.   Heavy breathing
        D.   Rolling the patient   Ref. 4 - p. 167

655.   To best examine the upper anterior ribs, the following projection is used:
        A.   Posterior oblique
        B.   Anterior oblique
        C.   Anteroposterior
        D.   Postero-anterior   Ref. 4 - p. 168

656.   In the AP projection of the ribs above the diaphragm, the film is centered
        at the level of the:
        A.   1st thoracic vertebra
        B.   6th thoracic vertebra
        C.   12th thoracic vertebra
        D.   1st lumbar vertebra   Ref. 4 - p. 169

657.   In the AP projection of the ribs below the diaphragm, the film is centered
        at the level of the:
        A.   1st thoracic vertebra
        B.   6th thoracic vertebra
        C.   12th thoracic vertebra
        D.   1st lumbar vertebra   Ref. 4 - p. 169

658.   The axillary portions of the ribs are best studied using either the anterior
        or posterior _____ projection:
        A.   Oblique
        B.   Prone
        C.   Supine
        D.   Recumbent   Ref. 4 - p. 170

659.   The hip bone consists of three parts: the ileum, ischium and:
        A.   Femur
        B.   Acetabulum
        C.   Trochanter
        D.   Pubis   Ref. 4 - p. 176

660. The joints where the ilia and sacrum articulate are called the:
    A. Ischio pubic
    B Sacroiliac
    C. Sciatic
    D. Innominate   Ref. 4 - p. 176

661. The 2 large bony processes found on the proximal femur are called the:
    A. Tuberosities
    B. Condyles
    C. Trochanters
    D. Epicondyles   Ref. 4 - p. 177

662. In the average adult, the neck and shaft of the femur form an angle of
     approximately:
    A. 90 degrees
    B. 120 degrees
    C. 180 degrees
    D. 240 degrees   Ref. 4 - p. 178

663. A palpable bony point to help localize the hip joint for positioning is the:
    A. Anterior superior iliac spine
    B. Sciatic notch
    C. Ischial spine
    D. Obturator foramen   Ref. 4 - p. 179

664. The symphysis pubis is at the same level as which other palpable bony
     landmark?:
    A. Ischial tuberosity
    B. Tip of the coccyx
    C. Iliac crest
    D. Greater trochanter of the femur   Ref. 4 - p. 180

665. In the antero-posterior projection of the hip, the foot should usually be
     slightly:
    A. Dorsiflexed
    B. Plantar flexed
    C. Everted
    D. Inverted   Ref. 4 - p. 182

666. In the cross table lateral projection for the femoral neck, the cassette is
     placed in contact with the:
    A. Lateral surface of the affected side
    B. Medial surface of the affected side
    C. Bottom of the affected side
    D. Top of the affected side   Ref. 4 - p. 190

667. For the anteroposterior projection of the iliac crest, the affected side is:
    A. Placed parallel to the film
    B. Elevated
    C. Abducted
    D. Placed perpendicular to the film   Ref. 4 - p. 206

668. For the posterior oblique position of the iliac crest, the patient is placed:
    A. Supine
    B. Prone
    C. Erect
    D. Lateral   Ref. 4 - p. 206

669. For the anteroposterior projection of the pubic and ischial rami, the
      central ray is angled:
      A. Toward the head
      B. Toward the feet
      C. Medially
      D. Laterally   Ref. 4 - p. 202

670. For the anteroposterior projection of the pelvis, the cassette is centered
      at the level of the:
      A. Top of the greater trochanter
      B. Top of the iliac crest
      C. Pubic symphysis
      D. Ischial tuberosity   Ref. 4 - p. 183

671. The patient is seated on the edge of the radiographic table, bending
      forward and grasping the ankles st that the pelvis is tilted forward. The
      central ray is directed vertically through the lumbosacral region. This
      projection is called:
      A. Chassard-Lapine
      B. Lateral
      C. Taylor position
      D. Lilienfeld position   Ref. 4 - p. 188

672. There are 7 cervical vertebrae and _____ dorsal vertebrae:
      A. Five
      B. Seven
      C. Twelve
      D. Fifteen   Ref. 4 - p. 208

673. The normal anteroposterior lumbar vertebrae curve is:
      A. Concave forward
      B. Lateral
      C. Convex backward
      D. Convex forward   Ref. 4 - p. 208

674. A typical vertebra is composed of 2 main parts: the body and the posterior
      bony ring called:
      A. Vertebral arch
      B. Coccyx
      C. Centrum
      D. Transverse process   Ref. 4 - p. 209

675. The first cervical vertebra is also called the:
      A. Lateral mass
      B. Atlas
      C. Axis
      D. Sacrum   Ref. 4 - p. 210

676. The conical process arising from the upper part of the body of C2 is the:
      A. Spinous process
      B. Transverse process
      C. Odontoid process
      D. Atlas   Ref. 4 - p. 211

677. In the AP projection for the atlas and axis, the patient is instructed to:
      A. Breathe
      B. Close the mouth
      C. Open the mouth widely
      D. Move from side to side   Ref. 4 - p. 220

678. For the lateral view of the axis and atlas, the central ray is directed to
       a point:
       A. 1 inch below the mastoid tip
       B. 1 inch below the mandible
       C. At the external acoustic meatus
       D. At the outer canthus of the eye   Ref. 4 - p. 225

679. In the AP projection of the cervical spine, the patient's chin should be
       extended so as to prevent:
       A. Superimposition of the mandible and the mid-cervical segments
       B. Motion
       C. Superimposition of the sternum and the mid-cervical segments
       D. Respiration   Ref. 4 - p. 226

680. To better define the intervertebral spaces in the AP projection of the
       cervical spine, the central ray is angled:
       A. 45 degrees toward the feet
       B. 15-20 degrees toward the feet
       C. 45 degrees toward the head
       D. 15-20 degrees toward the head   Ref. 4 - p. 226

681. To demonstrate the intervertebral foramina of the cervical spine, the
       following projection must be obtained:
       A. Erect AP
       B. Oblique
       C. Erect lateral
       D. Supine lateral   Ref. 4 - pp. 232, 235

682. To demonstrate the forward and backward movement of the cervical spine
       the following projections must be obtained:
       A. Open and closed mouth
       B. Erect and supine lateral
       C. Flexion and extension laterals
       D. Anterior and posterior obliques   Ref. 4 - p. 231

683. For the AP projection of the dorsal spine, the film is centered at the
       level of:
       A. C7
       B. D1
       C. D6
       D. D12   Ref. 4 - p. 240

684. In the routine lateral projection of the dorsal spine, the following
       structures are usually not well visualized:
       A. Upper 3-4 dorsal vertebrae
       B. The mid-dorsal vertebrae
       C. Lower spinous processes
       D. Mid-dorsal intervertebral disc spaces
                                                    Ref. 4 - p. 243

685. For the lateral projection of the lumbar spine, the film is centered at
       the level of the:
       A. Greater trochanter
       B. Pubic symphysis
       C. Iliac crest
       D. Sacral promontory   Ref. 4 - p. 250

686   For the oblique projection of the lumbar spine, the body is rotated about:
        A. 10 degrees
        B. 45 degrees
        C. 75 degrees
        D. 90 degrees   Ref. 4 - p. 255

687.   The oblique projection of the lumbar spine is used to demonstrate the:
        A. Spinous processes
        B. Transverse processes
        C. Apophyseal joints
        D. Upper sacrum   Ref. 4 - pp. 254, 255

688.   For the AP view of the sacrum and sacroiliac joints, the central ray
        is angled:
        A. 15-30 degrees caudad
        B. 15-30 degrees cephalad
        C. 60-75 degrees caudad
        D. 60-75 degrees cephalad   Ref. 4 - p. 262

689.   For the oblique view of the sacroiliac joints, the body is rotated:
        A. 5 degrees
        B. 25 degrees
        C. 60 degrees
        D. 75 degrees   Ref. 4 - p. 258

690.   For the lateral projection of the sacrum, the film is centered to the:
        A. Coccyx
        B. Xiphoid
        C. Anterior-superior iliac spines
        D. Symphysis pubis   Ref. 4 - p. 263

691.   For the AP projection of the coccyx, the central ray is angled:
        A. 35 degrees cephalad
        B. 60 degrees caudad
        C. 10 degrees cephalad
        D. 10 degrees caudad   Ref. 4 - p. 262

692.   In AP stress studies of the ankle _____ is used:
        A. Long exposure time
        B. Short exposure time
        C. Non-screen film   Ref. 4 - p. 91

693.   For the lateral projection of the skull, the central ray is directed through
        the sella turcica; this point can be found externally by centering:
        A. 1/2 inch above the outer canthus
        B. 3/4 inch anterior to and above the external auditory meatus
        C. 3/4 inch above the interpupillary line
        D. At the acantho-meatal line   Ref. 4 - p. 332

694.   In a properly centered lateral film of the skull, the:
        A. Mandibles will be superimposed
        B. Mandibles should not be superimposed
        D. Head is rotated 15 degrees   Ref. 4 - p. 333

695.   For the PA projection of the skull, the orbito-meatal line is placed
        perpendicular to the film and the central ray is angled:
        A. 15 degrees cephalad
        B. 35 degrees cephalad
        C. 15 degrees caudad
        D. 25 degrees caudad   Ref. 4 - p. 334

696. For the semi-axial AP projection of the skull, the chin may be depressed
     and the central ray tilted caudally to a combined total angle of:
     A. 5 degrees
     B. 15 degrees
     C. 40 degrees
     D. 75 degrees   Ref. 4 - p. 336

697. To project the dorsum sellae within the foramen magnum in a skull of
     average shape, the orbito-meatal line is placed perpendicular to the film
     and the central ray angled cranially:
     A. 5 degrees
     B. 15 degrees
     C. 25 degrees
     D. 60 degrees   Ref. 4 - p. 337

698. For the submentovertical projection (base view) of the skull, the central
     ray is directed at right angles to the:
     A. Infraorbito-meatal line
     B. Interpupillary line
     C. External auditory meatus
     D. Outer canthus   Ref. 4 - p. 340

699. If a non-bucky spot film of the sella turcica is requested, the following
     should be used:
     A. Adjustable angle block
     B. Extension cone
     C. Occlusal film
     D. Rapid film changer   Ref. 4 - p. 346

700. With the patient's head rotated 53 degrees, the orbit centered to the film
     and the central ray directed vertically to the mid-point of the film, one
     obtains a view of the:
     A. Sella turcica
     B. Sphenoid sinus
     C. Mastoid process
     D. Dorsum sellae   Ref. 4 - p. 352

701. In the lateral projection of the skull, the following structure is obscured
     by superimposed shadows of the opposite side of the skull:
     A. Sella turcica
     B. Sphenoid sinus
     C. Mastoid process
     D. Dorsum sellae   Ref. 4 - p. 378

702. For a stereoscopic examination of the mastoids, the tube is shifted:
     A. Transversely
     B. Up and down
     C. Obliquely
     D. Not at all   Ref. 4 - p. 380

703. In radiographing the mastoid processes:
     A. The patient's ears should be folded forward and taped
     B. The patient should be instructed to breathe continuously
     C. The head need not be immobilized
     D. A large focal spot size tube should be used
                                                      Ref. 4 - p. 381

704.   In the law (lateral) projection of the mastoids, the central ray is projected
        through a point 1 inch posterior to the external acoustic meatus and is
        angled:
        A. 25 degrees toward the back of the head
        B. Not angled at all
        C. 15 degrees toward the face and 15 degrees caudad
        D. 15 degrees toward the face and 15 degrees cephalad
                                                Ref. 4 - p. 382

705.   In the axio-lateral (Schüller) projection of the mastoids, the central ray
        is angled:
        A. 10 degrees caudad
        B. 10 degrees cephalad
        C. 25 degrees cephalad
        D. 25 degrees caudad   Ref. 4 - p. 390

706.   In the semi-axial view of the pars petrosa (Mayer's view), the head is:
        A. Not rotated
        B. Hyperextended
        C. Rotated 45 degrees toward the side examined
        D. Rotated 45 degrees away from the side examined
                                                Ref. 4 - pp. 400,401

707.   In the semi-axial view of the pars petrosa (Mayer's view), the central
        ray is directed:
        A. 45 degrees caudad
        B. 45 degrees cephalad
        C. 15 degrees caudad
        D. 15 degrees cephalad   Ref. 4 - pp. 400,401

708.   In the posterior profile projection of the mastoids (Stenver's view),
        the head:
        A. Is parallel to the film
        B. Forms a 45 degrees angle to the film
        C. Is perpendicular to the film
        D. Forms a 15 degrees angle to the film   Ref. 4 - p. 403

709.   In the Stenver's view of the mastoids, the central ray is angled:
        A. 12 degrees cephalad
        B. 45 degrees cephalad
        C. 12 degrees caudad
        D. 45 degrees caudad   Ref. 4 - p. 403

710.   A special mastoid apparatus and petrosa localizer is needed for the
        following mastoid projection:
        A. Low-Beer
        B. Granger
        C. Lysholm
        D. Arcelin   Ref. 4 - p. 384

711.   With the head resting on the forehead and nose, its median plane per-
        pendicular to the midline of the table, and the central ray angled 25
        degrees cephalad, one obtains which view of the petrous pyramids?:
        A. Granger
        B. Lysholm
        C. Haas
        D. Mayer   Ref. 4 - p. 394

712.   In the semi-axial projection for the petrous pyramids, the infra-
       orbitomeatal line is vertical, and the central ray is directed:
       A. Vertically
       B. 60 degrees caudally
       C. 30 degrees cephalad
       D. 30 degrees caudally   Ref. 4 - p. 392

713.   The semi-axial AP projection of the skull is excellent for visualizing the:
       A. Frontal sinuses
       B. Internal acoustic canals
       C. Optic foramina
       D. Carotid foramina   Ref. 4 - p. 393

714.   With the central ray directed at right angles to the orbitomeatal line,
       one obtains the following view of the petrous bones:
       A. Submento-vertical
       B. Occipito-frontal
       C. Posterior profile
       D. Lateral   Ref. 4 - p. 396

715.   For the semi-axial PA projection of the temporal styloid processes
       (Cahoon):
       A. Rest head on the forehead and nose and angle central ray 25 degrees
           cephalad
       B. Rest head on the forehead and nose and angle central ray 25 degrees
           caudad
       C. Rest the head on the chin and direct central ray at right angles to
           infraorbito-meatal line
       D. Rotate head 45 degrees away from the side examined and angle central
           ray 10 degrees caudad   Ref. 4 - p. 411

716.   The lateral view of the temporal styloid processes is taken with:
       A. The patient breathing
       B. The patient's mouth opened
       C. The patient's mouth closed
       D. Central ray angled 35 degrees caudad   Ref. 4 - p. 413

717.   In order to demonstrate the presence of fluid in the paranasal sinuses,
       the patient should be examined in which position?:
       A. Prone
       B. Supine
       C. Erect
       D. Lateral decubitus   Ref. 4 - p. 423

718.   The PA (Caldwell) projection of the skull is used to demonstrate the:
       A. Frontal and ethmoid sinuses
       B. Sphenoid sinuses
       C. Maxillary sinuses
       D. Optic foramina   Ref. 4 - p. 427

719.   The water's projection is used to demonstrate the:
       A. Mastoid tips
       B. Frontal sinuses
       C. Sphenoid sinuses
       D. Maxillary sinuses   Ref. 4 - p. 428

720. In the water's projection, the central ray is directed perpendicularly to
      the film, and the head is rested on the:
      A. Forehead
      B. Occiput
      C. Nose
      D. Chin   Ref. 4 - p. 428

721. The oblique position for the paranasal sinuses is also known as the:
      A. Fuch's position
      B. Valdini position
      C. Rhese position
      D. Water's position   Ref. 4 - p. 432

722   The vertico-submental projection is used to demonstrate which sinus
      group?:
      A. Sphenoid
      B. Frontal
      C. Maxillary
      D. Ethmoid   Ref. 4 - p. 435

723. To demonstrate the patella in a PA views of the knee exposure must be
      increased _____ kilovolts:
      A. 10
      B. 20
      C. 5
      D. Any of the above   Ref. 4 - p. 102

724. Measurement of lengths of long bones by radiographic methods is
      known as:
      A. Scanography
      B. Orthorography
      C. Kymography
      D. Arthrography   Ref. 4 - p. 110

725. To obtain a profile view of the glenoid fossa the central ray is _____
      and the opposite shoulder is elevated _____ degrees:
      A. Vertical, 5
      B. Oblique, 5
      C. Vertical, 10
      D. Vertical, 45   Ref. 4 - p. 129

726. With the head in a lateral position, the malar bone centered to the midline
      of the table, the cassette centered to the zygoma, and the central ray
      vertical, one obtains a lateral projection for the:
      A. Paranasal sinuses
      B. Mastoids
      C. Skull
      D. Facial bones   Ref. 4 - p. 440

727. A view used to demonstrate the facial bones, orbits and zygomatic
      arches is the:
      A. Water's
      B. Submento-vertical
      C. Apical lordotic
      D. Towne-Chamberlain   Ref. 4 - p. 442

728. Examination of the nasal bones should consist of which view(s)?:
      A. Oblique
      B. Lateral
      C. Base
      D. Lateral and occlusal   Ref. 4 - pp. 446, 448

729.   Medial or lateral displacement in a nasal bone fracture is best demon-
       strated by using which view?:
       A. Occlusal (axial)
       B. Lateral
       C. Base
       D. Lateral stereo   Ref. 4 - p. 448

730.   The following view can be used to demonstrate both zygomatic arches:
       A. Occlusal
       B. Submento-vertical
       C. Lateral
       D. Apical lordotic   Ref. 4 - p. 450

731.   To examine the zygomatic arches in the AP position the head is rotated:
       A. 15 degrees toward the side examined
       B. 15 degrees away from side examined
       C. 35 degrees toward the side examined
       D. 35 degrees away from the side examined
                                             Ref. 4 - p. 452

732.   To examine the hard palate, the following projection is used:
       A. Water's
       B. Intra-oral
       C. Oblique
       D. Chin down   Ref. 4 - p. 458

733.   For the oblique projection of the mandible, the central ray is angled:
       A. Cephalad
       B. Caudad
       C. Posteriorly
       D. Anteriorly   Ref. 4 - p. 464

734.   For a general PA survey film of the mandible, the central ray is:
       A. Perpendicular to the interpupillary line
       B. Perpendicular to the lips
       C. Angled 30 degrees cephalad
       D. Angled 30 degrees caudad   Ref. 4 - p. 470

735.   To demonstrate the rami and condyles of the mandible in the PA position,
       the central ray is:
       A. Perpendicular to the film
       B. Centered to the mandibular symphysis
       C. Angled 30 degrees cephalad
       D. Angled 30 degrees caudad   Ref. 4 - p. 471

736.   For the lateral projection of the temporo-mandibular joints, the film is
       centered to:
       A. A point 1/2 inch posterior to the external acoustic meatus
       B. A point 1/2 inch anterior to the external acoustic meatus
       C. The outer canthus of the eye
       D. The nasion   Ref. 4 - p. 478

737.   The following two exposures are generally taken when examining the
       temporo-mandicular joints:
       A. Inspiration and expiration
       B. AP and lateral
       C. Open and closed mouth
       D. Both obliques   Ref. 4 - p. 478

738. Periapical projections are used in roentgenography of the:
      A. Chest
      B. Teeth
      C. Brain
      D. Mastoids   Ref. 4 - p. 497

739. Roentgenographic demonstration of the salivary ducts by injection of
      contrast material is called:
      A. Sialography
      B. Placentography
      C. Planigraphy
      D. Salivation   Ref. 4 - p. 489

740. When examining the salivary glands, a _____ technique should
      be used:
      A. Overpenetrated Bucky
      B. Laminographic
      C. Barium swallow
      D. Soft tissue   Ref. 4 - p. 489

741. When examining the salivary glands, the following projections are
      commonly used:
      A. Both obliques
      B. Stereo AP and lateral
      C. Tangential, lateral and intra-oral
      D. PA, AP, and lateral   Ref. 4 - pp. 490,493

742. Soft tissue examination of the neck should include the following views:
      A. AP and lateral
      B. AP and PA
      C. Lateral in flexion and extension
      D. Both obliques   Ref. 4 - p. 529

743. In the Fisk position an axial view of_____ is obtained as well as a
      view of the bicipital groove:
      A. Glenoid fossa
      B. Lesser tuberosity
      C. Acromioclavicular joint
      D. Coracoid process   Ref. 4 - p. 130

744. To examine the costotransverse and costovertebral articulations the
      patient is placed in a supine position and the tube is angled _____
      degrees to the _____ :
      A. 20, head
      B. 15, feet
      C. 10, head
      D. 20, feet   Ref. 4 - p. 173

745. With the median sagittal plane of the skull and the orbitomeatal line
      perpendicular to the film and with the central ray angled 20-25 degrees
      to the feet a view of the _____ is obtained:
      A. Inferior orbital fissures
      B. Optic foramen
      C. Foramen magnum
      D. Superior orbital fissures   Ref. 4 - p. 360

746. The potential space between the lungs, within the thorax, is the:
      A. Diaphragm
      B. Trachea
      C. Mediastinum
      D. Larynx   Ref. 4 - p. 546

747. Diagnostic pneumoperitoneum studies require _____ in exposure
      technique:
      A. No change
      B. 10-20 kilovolt decrease
      C. 10-20 kilovolt increase
      D. 5 kilovolt decrease   Ref. 4 - p. 593

748. Gastric peristalsis is more active with the patient in the _____
      position:
      A. Right anterior oblique
      B. Left posterior oblique
      C. Supine
      D. Prone   Ref. 4 - p. 637

749. An organ that normally decreases in size with age is the:
      A. Thymus
      B. Thyroid
      C. Trachea
      D. Thorax   Ref. 4 - p. 550

750. For a standard roentgenogram of the chest, it is preferable to place the
      patient:
      A. Prone
      B. Supine
      C. Erect
      D. On his side   Ref. 4 - p. 551

751. In a PA roentgenogram of the chest, faulty positioning may result in
      rotation; such rotation can best be detected by comparing the position
      of the:
      A. Clavicles
      B. Dorsal vertebrae
      C. Diaphragm
      D. Trachea   Ref. 4 - p. 551

752. An exposure of the heart and lungs is usually made:
      A. In mid respiration
      B. During quiet breathing
      C. At the end of full inspiration
      D. At the end of full expiration   Ref. 4 - p. 552

753. To demonstrate the excursion of the diaphragm, the following films are
      obtained:
      A. Inspiration and expiration
      B. Anterior and posterior obliques
      C. Right and left lateral erect
      D. Both decubiti   Ref. 552

754. Stereoscopic projections of the chest are made with:
      A. Horizontal tube shifts
      B. Vertical tube shifts
      C. Laminography
      D. Oblique tube shifts   Ref. 4 - p. 552

755. In chest radiography the preferred focal-film distance is:
      A. 36 inches
      B. 40 inches
      C. 60 inches
      D. 72 inches   Ref. 4 - p. 553

756. In the erect lateral view of the chest, the patient's arms are placed:
    A. At the side
    B. On the hips
    C. Over the head
    D. On the cassette   Ref. 4 - p. 562

757. Examination of the chest for heart size and configuration is usually combined with:
    A. Laminograms
    B. Soft tissue examination of the neck
    C. Apical lordotic view
    D. Barium swallow   Ref. 4 - p. 565

758. With the patient erect, and the body rotated so that the left breast and shoulder are in contact with the cassette, the following projection of the chest is obtained:
    A. Left oblique
    B. Right oblique
    C. Left lateral
    D. Right lateral   Ref. 4 - p. 564

759. The patient stands erect and leans backward so as to rest the shoulders on the cassette. This view of the chest is called:
    A. Anterior oblique
    B. Anterior lordotic
    C. Posterior oblique
    D. AP   Ref. 4 - p. 568

760. The patient lies on his side with the cassette placed vertically against either the anterior or posterior surface of the chest. This view is the:
    A. Prone oblique
    B. Supine oblique
    C. Lateral decubitus
    D. Lordotic   Ref. 4 - p. 570

761. For the AP projection of the abdomen, the cassette is centered to the level of the:
    A. Pubic symphysis
    B. Iliac crests
    C. Xiphoid
    D. Greater trochanter   Ref. 4 - p. 588

762. In order to demonstrate air-fluid levels in the intestine or free air in the abdomen, the following view of the abdomen is used:
    A. Supine
    B. Prone
    C. Erect
    D. Lateral   Ref. 4 - p. 590

763. Cholecystography refers to the roentgenographic examination of the:
    A. Biliary system and gall bladder
    B. Colon
    C. Kidneys and ureters
    D. Urinary bladder   Ref. 4 - p. 598

764. Oral cholecystographic media produce a roentgen shadow because of their:
    A. Barium content
    B. Saline content
    C. Lead content
    D. Iodine content   Ref. 4 - p. 598

765.   Before administering the tablets for a cholecystogram, the following
        preliminary examination should be obtained:
        A.   Barium enema
        B.   Survey film of abdomen
        C.   Fatty meal
        D.   Upper GI series   Ref. 4 - p. 601

766.   If there is non-visualization on an oral cholecystogram, the following
        method is then frequently used:
        A.   Intravenous cholangiography
        B.   Intravenous urography
        C.   Nephrotomography
        D.   T-tube cholangiography   Ref. 4 - p. 606

767.   For the completion of the oral cholecystographic study, the patient
        is given:
        A.   An enema
        B.   A fatty meal
        C.   A laxative
        D.   A test dose of contrast material   Ref. 4 - p. 603

768.   Routine projections in oral cholecystography include the following views:
        A.   PA, oblique and erect
        B.   PA, AP and left posterior oblique
        C.   PA and left anterior oblique
        D.   PA and right posterior oblique   Ref. 4 - p. 611

769.   Examination of the upper gastro-intestinal tract is carried out using:
        A.   Erect films only
        B.   Both fluoroscopy and radiography
        C.   Horizontal films only
        D.   Fluoroscopy only   Ref. 4 - p. 626

770.   In examinations of the stomach, the exposure time should be no greater
        than:
        A.   3 seconds
        B.   2 seconds
        C.   1.5 seconds
        D.   0.5 seconds   Ref. 4 - p. 631

771.   The following positions are frequently used for examination of the
        esophagus:
        A.   PA, AP and erect
        B.   Supine and erect LAO
        C.   PA, RAO and lateral
        D.   Both lateral decubiti   Ref. 4 - p. 633

772.   For an upper gastro-intestinal series, the patient is allowed:
        A.   Nothing by mouth for 24 hours prior to the examination
        B.   Nothing by mouth for 6 hours prior to the examination
        C.   A full diet
        D.   Nothing by mouth for 30 minutes prior to the examination
                                                      Ref. 4 - p. 635

773.   For the recumbent oblique view of the stomach and duodenum the body is
        rotated:
        A.   5-10 degrees
        B.   20-35 degrees
        C.   40-70 degrees
        D.   75-90 degrees   Ref. 4 - p. 644

774. After oral ingestion of barium, films of the abdomen are made at
      30-60 minute intervals if an examination of the _____ is desired:
      A. Small bowel
      B. Esophagus
      C. Stomach
      D. Gall bladder   Ref. 4 - p. 653

775. A double contrast enema consists of:
      A. Pre- and post evacuation films of the colon
      B. Air and barium examination of the colon
      C. Air and barium examination of the stomach
      D. Refilling the colon with barium   Ref. 4 - pp. 668,669

776. The double contrast enema is useful to demonstrate:
      A. Polyps of the colon
      B. The rectum
      C. The small bowel
      D. The stomach   Ref. 4 - p. 656

777. The following projection is useful to visualize the recto-sigmoid area:
      A. Lateral decubitus
      B. Post-evacuation
      C. Double contrast
      D. Chassard-Lapine   Ref. 4 - p. 667

778. The intravenous injection of contrast material to visualize the urinary
      system is called:
      A. Excretory urography
      B. Retrograde pyelography
      C. Voiding urethrography
      D. Intravenous cholangiography   Ref. 4 - p. 686

779. Prior to the intravenous method of urinary tract visualization, the patient:
      A. Is not allowed to drink liquids
      B. Drinks 5 glassfuls of water
      C. Is never given enemas or laxatives
      D. Is given a large meal   Ref. 4 - p. 692

780. Prior to the radiologic investigation of the urinary tract, the following
      examination is obtained:
      A. Retrograde pyelogram
      B. Survey film of abdomen
      C. Barium enema
      D. Esophagogram   Ref. 4 - p. 695

781. Radiography of the breast is called:
      A. Pelvimetry
      B. Mamillation
      C. Placentography
      D. Mammography   Ref. 4 - p. 726

782. Examination of the uterus and fallopian tubes by the instillation of
      contrast material is:
      A. Hysterosalpingography
      B. Fetography
      C. Placentography
      D. Urethrography   Ref. 4 - p. 744

783.   The roentgen study used to compare the size of the maternal pelvis and
        the head of the fetus is called:
        A. Placentography
        B. Pelvimetry
        C. Hysterography
        D. Planigraphy   Ref. 4 - p. 752

784.   A view of the pelvic outlet is obtained using the:
        A. Colcher-Sussman method
        B. Caldwell-Molloy method
        C. Ball method
        D. Chassard-Lapine position   Ref. 4 - p. 772

785.   For the above view of the pelvic outlet, the central ray is directed
        vertically to the median plane of the:
        A. Sacro-coccygeal junction
        B. Sacrum, at the level of the highest point of the greater trochanters
        C. Pubic symphysis
        D. Iliac spines   Ref. 4 - p. 772

        INDICATE WHETHER EACH OF THE FOLLOWING STATEMENTS IS
        (T)RUE OR (F)ALSE:

786.   To insure visualization of C7 in lateral view of the cervical spine the
        central ray should be directed cephalad.   Ref. 15 - p. 73

787.   The right intervertebral foramina of the cervical spine are visualized
        when films are obtained with the patient in the left anterior oblique position.
                                                Ref. 15 - p. 73

788.   In the Stenver's position for study of the temporal bone the sagittal plane
        of the skull makes a 45 degree angle with the film.
                                                Ref. 15 - p. 113

789.   The central ray is angled 12 degrees toward the feet in the Stenver's
        projection of the temporal bone.   Ref. 15 - p. 113

790.   In examination of the elbow in the lateral projection the epicondyles of
        the humerus must be exactly superimposed and perpendicular to the film.
                                                Ref. 15 - p. 517

791.   The Towne's projection of the skull gives a good view of the occipital bone
        and foramen magnum.   Ref. 15 - p. 5

792.   In the submento vertical projection of the skull, the cantho-meatal line is
        perpendicular to the film.   Ref. 15 - p. 6

793.   No more than AP and lateral views are required in knee arthrography.
                                                Ref. 15 - p. 511

794.   A complete study of the temporo-mandibular joints includes views
        obtained with the mouth open and closed.   Ref. 15 - p. 12

795.   For a lateral view of the mandible the head is in a true lateral position
        and the tube is angled 25-30 degrees cephalad.
                                                Ref. 15 - p. 8

796.   Laminography is a valuable method of examining facial bones in patients
        who cannot be moved because of injury. Ref. 15 - p. 12

797.   Oblique radiographs are valuable in cervical myelography but not in
       lumbar myelography.   Ref. 15 - p. 83

798.   In the AP projection of the sacroiliac joints the tube is angulated toward
       the head.   Ref. 15 - p. 78

799.   A preliminary scout film is an absolute necessity in any urographic
       procedure   Ref. 15 - p. 249

800.   In order to superimpose the femoral condyles in the lateral view of the
       knee the central ray must be perpendicular to the film.
                                                  Ref. 15 - p. 501

801.   Stress studies of the ankle joint are done in the lateral position.
                                                  Ref. 15 - p. 504

802.   The conventional PA projection of the hand presents an oblique view of
       the thumb.   Ref. 15 - p. 518

803.   A good PA view of the carpal navicular is obtained with the hand in ulnar
       deviation.   Ref. 15 - p. 518

804.   Caldwell projection of the skull is a PA view with 15-20 degrees of caudal
       tilt of the tube.   Ref. 15 - p. 5

805.   The Towne's projection represents a PA view of the skull with 30-45
       degree cephalad tube tilt.   Ref. 15 - p. 5

806.   In the Water's position the orbito-meatal line is perpendicular to the film.
                                                  Ref. 15 - p. 105

807.   The Water's position with the open mouth is used to demonstrate the
       sphenoid sinus.   Ref. 15 - p. 105

808.   A lateral view of the mastoid air cells is obtained with the Law projection.
                                                  Ref. 15 - p. 114

809.   In the Law projection the central ray is directed 15 degrees cephalad and
       15 degrees posteriorly.   Ref. 15 - p. 114

810.   The Schüller projection is exposed with the head in a true lateral position
       with the tube angled 25 degrees caudad.   Ref. 15 - p. 114

811.   The radio-carpal articulation is best shown with the wrist in the PA
       position and the tube angled 25 to 30 degrees toward the elbow joint.
                                                  Ref. 7 - p. 19

812.   The AP view of the elbow demonstrates the circumference of the radial
       head.   Ref. 7 - p. 36

813.   The glenohumeral joint is demonstrated best in AP position with the
       opposite shoulder rotated 45 degrees away from the film.
                                                  Ref. 7 - p. 56

814.   When a fracture of the calcaneus is suspected the axial view should be
       obtained.   Ref. 7 - p. 93

815.   The sternal angle is at the level of the T4-T5 interspace.
                                                  Ref. 7 - p. 175

816.   The umbilicus is at the level of L1.   Ref. 7 - p. 175

817.   The radiographic baseline and the orbitomeatal line are identical.
        Ref. 7 - p. 253

818.   When the interorbital line is perpendicular to the film, the head is
        properly positioned for a true lateral view of the skull.
        Ref. 7 - p. 253

819.   The best views of the optic foramina are obtained with the orbitomeatal
        line perpendicular to the film.   Ref. 7 - p. 369

820.   Positioning the head for the Stenver's projection places the petrous
        temporal bone parallel to the film.   Ref. 7 - p. 389

821.   Examination of the urinary tract is most frequently performed with the
        patient in oblique or upright position.   Ref. 7 - p. 541

822.   Gynecography is performed with the patient prone and the table tilted so
        that the head is lower than the feet.   Ref. 7 - p. 577

823.   In the PA projection of the abdomen the kidneys appear to lie in a more
        superior position than in the AP projection.
        Ref. 7 - p. 546

824.   In the PA projection of the hand centering is over the 2nd proximal
        phalanx.   Ref. 7 - p. 4

825.   Individual metacarpal bones are shown to best advantage in the lateral view.
        Ref. 7 - p. 4

826.   In examination of the shoulder girdle respiratory motion causes no
        difficulty.   Ref. 7 - p. 42

827.   The lateral projection of the scapula may be obtained with patient upright
        or prone.   Ref. 7 - p. 62

828.   With the patient prone and one shoulder elevated twelve inches from the
        table top, the opposite scapula lies parallel to the table top.
        Ref. 7 - p. 62

829.   The articulation of the talus and calcaneus are best shown in the AP
        projection.   Ref. 7 - p. 96

830.   The centering point for a lateral view of the ankle is the cuboid bone.
        Ref. 7 - p. 100

831.   Instability of the ankle joint cannot be determined with routine AP and
        lateral views.   Ref. 7 - p. 102

832.   The routine examination of the knee includes oblique views.
        Ref. 7 - p. 115

833.   When the lower extremity is externally rotated, the femoral neck appears
        to be foreshortened.   Ref. 7 - p. 128

834.   Landmarks used in localizing the head of the femur are the greater trochanter of the femur and the anterior superior iliac spine.

Ref. 7 - p. 131

835.   With the patient in a supine position the symphysis pubis and the anterior superior iliac spine are always in the same horizontal plane.

Ref. 7 - p. 154

FOR EACH OF THE FOLLOWING MULTIPLE CHOICE QUESTIONS
SELECT THE ONE MOST APPROPRIATE ANSWER:

836.   Photographic fog may be caused by:
        A. Film too old
        B. Improper safelight
        C. Radiation exposure
        D. All of the above   Ref. 13 - p. 243

837.   Streakiness of film is caused by:
        A. Failure to agitate film during development
        B. Film left too long in developer
        C. Sudden changes of temperature during processing
        D. Developer too warm   Ref. 13 - p. 244

838.   Reticulation (spidery marks) are due to:
        A. Intereaction of developer and fixer
        B. Faulty composition of developer
        C. Air bubbles on the film
        D. Sudden changes in temperature during processing
                                          Ref. 13 - p. 245

839.   In automatic processors, one of the following is not due to improperly
        seated transport rollers:
        A. Failure of film to transport
        B. Surface scratches or plus density lines on film
        C. Abnormal film densities   Ref. 12

840.   Plus density lines running in direction of film travel are due to:
        A. Missing or improperly seated transport rollers
        B. Contaminated developer solution
        C. Empty replenisher tanks
        D. Wash water dirty   Ref. 12

841.   Plus density lines regularly spaced perpendicular to direction of film
        are due to:
        A. Turn around out of adjustment
        B. Exhausted developer solution
        C. Abnormal roller surface
        D. Developer contaminated by fixer   Ref. 12

842.   Plus density lines irregularly spaced perpendicular to the film travel,
        are due to:
        A. Transport roller not rotating
        B. Turn about out of adjustment
        C. Loose rack chain causing rollers to hesitate
        D. Wash water dirty   Ref. 12

843.   Frilling (loosening of film emulsion from its base) occurs when:
        A. Film contains dried-in residue of fixer
        B. Developer is too warm
        C. Sudden changes of temperature during processing
        D. All of the above   Ref. 13 - p. 245

844.   Phenidon:
        A. Is a strong developing agent
        B. Replaces metol (elon)
        C. In combination with Hydroquinone gives a developer producing more
            contrast
        D. All of the above   Ref. 13 - p. 226

845.   The temperature of the developer lies in general between:
          A. 65 degrees F and 70 degrees F
          B. 68 degrees F and 80 degrees F
          C. 65 degrees F and 75 degrees F
          D. 75 degrees F and 95 degrees F   Ref. 13 - p. 226

846.   The replenisher is so constituted as to correct the developer for:
       A. The decrease of sodium carbonate
       B. The increase of bromine contents
       C. The dilution of the reducers
       D. All of the above   Ref. 13 - p. 227

847.   X-ray film emulsion consists of:
       A. Metallic silver crystals
       B. Silver bromide crystals
       C. Potassium bromide crystals
       D. Silver nitrate crystals   Ref. 5 - p. 14

848.   The support for the emulsion is called the:
       A. Base
       B. Screen
       C. Film holder
       D. Negative   Ref. 5 - p. 14

849.   Chemically, gelatin is a(an):
       A. Compound
       B. Element
       C. Gas
       D. Colloid   Ref. 5 - p. 16

850.   Absorption of X-rays by matter depends on:
       A. Atomic number of the substance
       B. The density or concentration of the substance
       C. The thickness of the substance
       D. All of the above   Ref. 13 - p. 50

851.   Screen type film is particularly sensitive to:
       A. Direct action of X-rays
       B. Ultra violet light
       C. Fluorescent light
       D. Infra red light   Ref. 5 - p. 18

852.   Direct exposure type film is particularly sensitive to:
       A. Direct action of X-rays
       B. Ultra violet light
       C. Fluorescent light
       D. Infra red light   Ref. 5 - p. 18

853.   The speed of screen type film is _____ direct exposure type film:
       A. The same as
       B. Much faster than
       C. Much slower than   Ref. 5 - p. 18

854.   If screen type film is used without an intensifying screen, its speed is
       _____ the direct exposure type:
       A. The same as
       B. Slower than
       C. Faster than   Ref. 5 - p. 18

855.   Screen type film can be used without an intensifying screen when radiographing thin parts of the body in order to obtain:
       A.   A faster speed
       B.   Less detail
       C.   A wider range of density
       D.   A narrower range of density   Ref. 5 - p. 18

856.   Direct exposure film is used only for:
       A.   Thin body parts
       C.   Over penetrated bucky exposures
       D.   Portable films   Ref. 5 - p. 18

857.   The inherent contrast of direct exposure film as compared to directly exposed screen type film is:
       A.   Darker
       B.   The same as
       C.   Lighter   Ref. 5 - p. 19

858.   The effective atomic number of the soft parts of the human body is equal to that of:
       A.   Calcium
       B.   Water
       C.   Sodium chloride
       D.   Hydrogen   Ref. 13 - p. 72

859.   Direct exposure film cannot be used with:
       A.   Cardboard holders
       B.   Processing hangers
       C.   Automatic processors
       D.   Intensifying screens   Ref. 5 - p. 19

860.   Periapical refers to:
       A.   Film for use in the operating room
       B.   Dental X-ray film
       C.   Film for foreign bodies
       D.   Film for mammography   Ref. 5 - p. 19

861.   X-rays striking the crystals in the film emulsion produce a:
       A.   Latent image
       B.   Visual image
       C.   Virtual image
       D.   Reversed image   Ref. 5 - p. 22

862.   The whole set of contrasts contained in the emergent beam (remnant) results in a shadow pattern of the subject, called:
       A.   Latent image
       B.   Radiation image
       C.   Virtual image
       D.   None of the above   Ref. 13 - p. 73

863.   The purpose of the developer is to convert the image produced by X-ray energy absorption into a:
       A.   Latent image
       B.   Visible image
       C.   Virtual image
       D.   Reversed image   Ref. 5 - p. 24

864.   The developing solution converts the exposed crystals to:
        A. Silver bromide
        B. Potassium bromide
        C. Metallic silver
        D. Silver nitrate   Ref. 5 - p. 24

865.   The developing solution is a(an):
        A. Reducing agent
        B. Oxidizing
        C. Absorbing agent
        D. Wetting agent   Ref. 5 - p. 25

866.   During development, the developing agent becomes:
        A. Reduced
        B. Oxidized
        C. Diluted
        D. Replenished   Ref. 5 - p. 25

867.   The invisible radiation image can be rendered visible by:
        A. Fluoroscopy screens
        B. Recording on film
        C. Video tape recording
        D. All of the above   Ref. 13 - p. 74

868.   The phenomenon in which all images in a particular projection fall one
        above the other is called:
        A. Magnification
        B. Superimposition
        C. Distortion
        D. Composite   Ref. 13 - p. 77

869.   Paralactic shift is used to:
        A. Separate superimposed details
        B. Minimize distortion
        C. Increase magnification
        D. All of the above   Ref. 13 - p. 77

870.   When the X-ray beam is parallel to the long axis of a vessel, the
        rounded image on the film is due to:
        A. The paralax
        B. The magnification
        C. The "end on" effect
        D. Distortion   Ref. 13 - p. 80

871.   The ratio focus film distance over focus object distance represents:
        A. The paralax
        B. The end on effect
        C. The magnification ratio
        D. All of the above   Ref. 13 - p. 81

872.   Distortion is the greatest when:
        A. OFD is greatest
        B. OFD is smallest
        C. FFD is greatest
        D. When non-screen film is used   Ref. 13 - p. 82

873. In zonography the paning angle is usually equal to:
     A. 30 degrees
     B. 12 degrees
     C. 6 degrees
     D. 24 degrees   Ref. 13 - p. 99

874. In pantomography the object and film are rotated in:
     A. Opposite directions
     B. The same direction
     C. At different speeds
     D. None of the above   Ref. 13 - p. 104

875. Radiological magnification can improve:
     A. Definition
     B. Contrast
     C. Make details already visible much more distinct
     D. All of the above   Ref. 13 - p. 105

876. With a 0.3 mm focus the optimum magnification is about:
     A. X 3
     B. X 4
     C. X 2
     D. X 1.5   Ref. 13 - p. 106

877. The limitation of radiographic material to show detail is called:
     A. Definition
     B. Contrast
     C. Fog
     D. Intrinsic unsharpness   Ref. 13 - p. 150

878. The threshold of perception of detail depends upon:
     A. The quality of the detail
     B. The degree of adaptation of the eye
     C. The visual power of the eye
     D. All of the above   Ref. 13 - p. 152

879. One of the following is not a factor determining the quality of detail:
     A. Size of detail
     B. Degree of detail contrast
     C. Degree of detail unsharpness
     D. Patient's pathology   Ref. 13 - p. 157

880. At low light intesities we perceive an image by using certain cells
     located in the retina:
     A. Rods
     B. Cones
     C. Macula lutea
     D. Iris   Ref. 13 - p. 157

881. The modulation transfer function is used for:
     A. Expressing the relationship between Kv and Mas
     B. Expressing image quality
     C. Measuring the grid ratio
     D. Regulating the filament current   Ref. 13 - p. 158

882. Geometric unsharpness is inversely proportional to:
     A. The size of the focus
     B. The distance between object and film
     C. The distance between focus and film
     D. All of the above   Ref. 13 - p. 165

883. Movement unsharpness is due to:
    A. Focus movement
    B. Object movement
    C. Film movement
    D. All of the above   Ref. 13 - p. 167

884. Movement unsharpness is most severe:
    A. When objects move in the direction of the X-ray beam
    B. In movements parallel to film with object near film
    C. In movements parallel to film will object away from film
    D. None of the above   Ref. 13 - p. 167

885. Unsharpness caused by intensifying screens is:
    A. Intrinsic to the type of screens
    B. Smallest with high speed screens
    C. Dependent on the focus film distance
    D. Always constant for any type of screens
                                           Ref. 13 - p. 169

886. Unsharpness caused by the cassette can be due to:
    A. Bad screen contact with film
    B. Wear of felt in cassette
    C. Faulty fasteners
    D. All of the above   Ref. 13 - p. 170

887. The total unsharpness is smallest when:
    A. The individual unsharpnesses are a minimum
    B. The individual unsharpnesses are equal to each other
    C. One individual unsharpness is at a minimum
    D. All of the above   Ref. 13 - p. 171

888. Direct exposure medical film compared to the screen type:
    A. Has a thicker emulsion
    B. Has a thinner emulsion
    C. Has the same emulsion
    D. Reacts in the same manner to exposure to light
                                           Ref. 14 - p. 4

889. High speed film:
    A. Is less susceptible to background radiation
    B. Ages poorly
    C. Has higher contrast than medium speed films
    D. Has finer grain than medium speed films
                                           Ref. 14 - p. 7

890. Film quality in high speed film is vastly improved by:
    A. Using a 20 watt bulb in the safelight
    B. Automatic processing
    C. Sight development
    D. Using high speed intensifying screens
                                           Ref. 14 - p. 9

891. The phototiming cassette has:
    A. No lead on the back
    B. A thinner lead foil on the back
    C. The same lead foil as a conventional cassette
    D. A thicker lead foil   Ref. 14 - p. 9

892. Intensifying screens glow _____ when activated by radiation:
    A. Green
    B. Blue
    C. Blue violet
    D. White   Ref. 14 - p. 9

893. The front intensifying screen:
    A. Can be somewhat thinner than the back screen
    B. Is thicker than the back screens
    C. Is always the same as the back screen
    D. None of the above   Ref. 14 - p. 11

894. The most common method used to produce a slow speed or detail screen
    is to:
    A. Increase the layer of crystals
    B. Increase the size of the crystals
    C. Use the screen staining process
    D. Use a plastic base   Ref. 14 - p. 11

895. Screen resolution is expressed:
    A. In terms of lines per millimeters
    B. By comparison with another screen
    C. In relation to non-screen technique
    D. Cannot be expressed   Ref. 14 - p. 13

896. A distinct advantage of direct exposure technique is that:
    A. It exhibits extreme contrast
    B. Allows for less radiation to patient
    C. Offers a wide range of exposure latitude
    D. Is easier to process   Ref. 14 - p. 13

897. The major disadvantage in using graded screens is that:
    A. They are not universal
    B. They are difficult to use
    C. The patient is exposed to the full effect of the X-ray beam
    D. Exhibit unequal densities   Ref. 14 - p. 20

898. Fog is more detrimental to:
    A. The densest parts of the film
    B. The middle tones
    C. The areas of low density
    D. All of the above   Ref. 14 - p. 26

899. Fog-like image on finished radiographs are produced by:
    A. Scattered radiation
    B. Improper safelight
    C. Storage conditions
    D. All of the above   Ref. 14 - p. 26

900. One of the following is not a primary ray restricting device:
    A. Lead diaphragm
    B. Cone
    C. Potter-Bucky
    D. Collimator   Ref. 14 - p. 27

901.   The air gap technique is used:
        A. In direct magnification
        B. To reduce scatter
        C. In tomography
        D. In pantography   Ref. 14 - p. 27

902.   In collimation when the field size is larger than the film size, the gonadal
        dosage is from _____ times higher:
        A. 10 to 20
        B. 80 to 400
        C. 50 to 100
        D. 200 to 1000   Ref. 14 - p. 37

903.   The cross table motion of the grid results in a blurring of the grid lines:
        A. Parallel to the width of the table
        B. Oblique to the length of the table
        C. Parallel to the length of the table
        D. None of the above   Ref. 14 - p. 37

904.   Grids have definite _____ limitations:
        A. Milliamperage
        B. Kilovoltage
        C. Resistance
        D. Reactance   Ref. 14 - p. 39

INDICATE WHETHER EACH OF THE FOLLOWING STATEMENTS IS
(T)RUE OR (F)ALSE:

905.   In converting from non-grid technique to grid technique an increase in
        milliamperage is advised.   Ref. 14 - p. 39

906.   In converting grid to non-grid technique the conversion should be made
        by reducing the KV value.   Ref. 14 - p. 39

MULTIPLE CHOICE. SELECT THE MOST APPROPRIATE ANSWER:

907.   The central ray must always be perpendicular to:
        A. Parallel grid
        B. Focused grid
        C. Stationary grid
        D. Cross hatch grid   Ref. 14 - p. 43

908.   The central ray must always be central in:
        A. Parallel grids
        B. Focused grids
        C. Cross hatch grids
        D. All of the above   Ref. 14 - p. 51

909.   A fluoro spot film mask is used to:
        A. Reduce scatter
        B. Subdivide cassettes
        C. To protect the technologist
        D. All of the above   Ref. 14 - p. 57

910.   Conventional rotating anodes are _____ in diameter:
        A. 4 inches
        B. 7 inches
        C. 3 inches
        D. 2 inches   Ref. 14 - p. 74

911. The added filtration filters in diagnostic tubes:
A. Should be difficult to remove
B. May be easily removed
C. Are useless in diagnostic radiography
D. Are made of lead   Ref. 14 - p. 75

912. Variations in patient anatomy can be compensated by:
A. Decreasing KV
B. Increasing MAS
C. Using compensatory filters
D. Varying FFD   Ref. 14 - p. 77

913. Compensatory filters are used:
A. At the tube side (portal) of patient
B. Below the patient
C. Both A and B
D. None of the above   Ref. 14 - p. 78

914. A "fractional-focus" spot measures:
A. 1mm
B. .6mm
C. 0.3mm or less
D. 2mm   Ref. 14 - p. 80

915. Early magnification tubes had:
A. Loo large focus spots
B. Severe rating restrictions
C. Were too bulky
D. Stationary anodes   Ref. 14 - p. 80

916. When the object to be examined is placed midway between tube and film the resulting magnification is:
A. 2 X linear
B. 2-1/2 X linear
C. 4 X area
D. A and C   Ref. 14 - p. 81

917. Spring driven mechanical timers:
A. Operate often below 1/10 of a second
B. Rarely operates at 1/10 of a second or below
C. Is never used with portable units
D. Is used for pediatric radiography   Ref. 14 - p. 98

918. Synchronous timers cannot be relied on for exposures shorter than:
A. 1/20 of a second
B. 1. 10 of a second
C. 1/2 of a second
D. 1/100 of a second   Ref. 14 - p. 99

919. Some modern timers can control exposures that are of a duration of:
A. One impulse
B. Three impulses
C. A fraction of an impulse
D. Ten impulses   Ref. 14 - p. 99

920.   A spinning top test for 1/20 of a second for full wave rectification will
        show _____ dots on the film:
        A. 9
        B. 12
        C. 6
        D. 3   Ref. 14 - p. 101

921.   A spinning top test for 1/20 of a second in half wave rectification will
        show _____ dots on the film:
        A. 9
        B. 12
        C. 6
        D. 3   Ref. 14 - p. 101

922.   The conventional spinning top:
        A. Can be used for any type of rectification
        B. Can only be used for halfwave rectification
        C. Cannot be used in three phase rectification
        D. Is not used anymore   Ref. 14 - p. 101

923.   A complex device that reproduces radiographic exposures of consistent
        densities is called:
        A. Photo roentgen
        B. A serial device
        C. A phototimer
        D. A programmer   Ref. 14 - p. 101

924.   A phototimer:
        A. Corrects error
        B. Can be used by unskilled personnel
        C. Is not always accurate
        D. Does not negate the skill of the technologist
                                                      Ref. 14 - p. 101

925.   The phototimer is located:
        A. Between patient and Bucky
        B. Behind the cassette
        C. Between film and patient
        D. B and C   Ref. 14 - p. 103

926.   Minimal reaction time of a phototimer:
        A. Is the shortest exposure time obtainable
        B. Is independent of quality and quantity of X-ray beam
        C. May lead to over-exposure of thin patients
        D. All of the above   Ref. 14 - p. 105

927.   The "back up time" of a phototimer is:
        A. The maximum permissible time
        B. An additional timer in case of under-exposure
        C. The duration of exposure with a phototimer
        D. None of the above   Ref. 14 - p. 106

928.   A major consideration in formulating an X-ray technique should be the:
        A. Tube ratings
        B. Recommendations of the radiologist
        C. Control of radiation dosage to both patient and the operator
        D. Limitations of the equipment   Ref. 14 - p. 112

929. One of the following is inaccurate:
     A. Avoid multiple exposures
     B. Females in child bearing age must be questioned about their last menstrual period
     C. Use the lowest KV possible
     D. Mechanical restraining devices should be used whenever possible
                                          Ref. 14 - p. 113

930. Technique charts should be formulated:
     A. With awareness of KV limitation of the grid
     B. Are always applicable to any patient
     C. Must be corrected to suit the pathology of the patient
     D. A and C   Ref. 14 - p. 113

931. The use of _____ helps to eliminate technical variations from one patient to the other:
     A. Fixed KV technique charts
     B. Fixed MAS technique charts
     C. Phototimer
     D. Potter-Bucky   Ref. 14 - p. 114

932. In the range between 60 and 85 KVP, an increase of _____ KVP requires a 50% reduction in MAS:
     A. 4 KVP
     B. 10 KVP
     C. 15 KVP
     D. 7 KVP   Ref. 14 - p. 117

933. The MAS value of any given technique should be doubled or reduced by one half for every _____ centimeter of tissue variation up to 30 centimeters:
     A. 10
     B. 7
     C. 3
     D. 2   Ref. 14 - p. 117

934. In general a _____ scale contrast technique results in radiographs of more consistent diagnostic quality:
     A. Short scale
     B. Long scale
     C. Extra long scale
     D. All of the above   Ref. 14 - p. 118

(T)RUE OR (F)ALSE:

935. A simple but accurate method of determining processor accuracy is not necessary.   Ref. 14 - p. 113

936. Proper calibration of the radiographic unit is a must.
                                          Ref. 14 - p. 113

937. With some exceptions, technical factors should be selected prior to final positioning of the patient.   Ref. 14 - p. 113

938. A good technical history is vital to the making of a good quality radiograph.   Ref. 14 - p. 113

939. A seemingly minor move in the FFD (4 to 5 inches) does not influence density drastically.   Ref. 14 - p. 115

940.   For every body part there is an optimal Kvp value.
                                          Ref. 14 - p. 117

941.   Almost all contrast studies use low or moderate Kvp value.
                                          Ref. 14 - p. 118

942.   Studies utilizing air or gas as contrast agent require high Kv techniques.
                                          Ref. 14 - p. 120

943.   In Bucky chests, High Kv ranges should be utilized.
                                          Ref. 14 - p. 121

944.   Molybdenum targets and molybdenum filters produce a monochromatic
        beam.   Ref. 14 - p. 125

MULTIPLE CHOICE.   SELECT THE MOST APPROPRIATE ANSWER:

945.   Small mobile units have a _____ to _____ mA and _____
        to _____ Kvp limitation:
        A.   10 to 20 mA and 50 to 70 Kvp
        B.   15 to 30 mA and 85 to 90 Kvp
        C.   50 to 100 mA and 75 to 95 Kvp
        D.   80 to 100 mA and 90 to 100 Kvp   Ref. 14 - p. 127

946.   Nickel cadmium cell battery powered units are:
        A.   Less powerful than capicator discharge units
        B.   Have the same output than A
        C.   Similar in output as 3 phase units with 12 pulse rectification
        D.   None of the above   Ref. 14 - p. 129

947.   The most critical problem in bedside radiographic technique is:
        A.   Kvp limitations
        B.   mAs limitations
        C.   Variation in FFD
        D.   Weight of the machine   Ref. 14 - p. 130

948.   Bedside radiographs exhibit more _____ than conventional radiographs:
        A.   Density
        B.   Scatter
        C.   Contrast
        D.   Visibility of detail   Ref. 14 - p. 133

949.   When frequent portable films have to be taken of a femur in traction:
        A.   A grid should always be used
        B.   A grid should never be used
        C.   Use of grid should be minimal
        D.   None of the above   Ref. 14 - p. 133

(T)RUE OR (F)ALSE:

In portable radiography:

950.   The length of the switch cord is an important factor.
                                          Ref. 14 - p. 135

951.   The use of a lead apron is not recommended.
                                          Ref. 14 - p. 136

952.   Hazard proof units should be used in the O. R.
                                        Ref. 14 - p. 137

953.   A portable unit can be used for fluoroscopy.
                                        Ref. 14 - p. 137

954.   Additional radiographs of varying techniques should be made in remote
        locations.   Ref. 14 - p. 137

MULTIPLE CHOICE. SELECT THE MOST APPROPRIATE ANSWER:

955.   The selectivity of a grid is:
        A. The radiation contrast with grid to the contrast without grid
        B. The ratio of the percentage of transmitted primary to the percentage
            of transmitted scatter
        C. Ratio of exposure time with grid over exposure time without grid
        D. The ratio of the height of the strips to the spacing between them
                                        Ref. 13 - p. 182

956.   The contrast improvement factor of a grid is:
        A. The radiation contrast with grid to the contrast without grid
        B. The ratio of the percentage of transmitted primary to the percentage
            of transmitted scatter
        C. Ratio of exposure time with grid over exposure time without grid
        D. The ratio of the height of the strips to the spacing between them
                                        Ref. 13 - p. 183

957.   The grid factor of a grid is:
        A. The radiation contrast with grid to the contrast without grid
        B. The ratio of the percentage of transmitted primary to the percentage
            of transmitted scatter
        C. Ratio of exposure time with grid over exposure time without grid
        D. The ratio of the height of the strips to the spacing between them
                                        Ref. 13 - p. 183

958.   The grid ratio of a grid is:
        A. The radiation contrast with grid to the contrast without grid
        B. The ratio of the percentage of transmitted primary to the percentage
            of transmitted scatter
        C. Ratio of exposure time with grid over exposure time without grid
        D. The ratio of the height of the strips to the spacing between them
                                        Ref. 13 - p. 183

959.   By using a film cassette combination, the film contrast is _____
        times greater than the original radiation contrast:
        A. 3 to 4 times
        B. 1 to 2 times
        C. 2 to 3 times
        D. Not changed   Ref. 13 - p. 186

960.   The radiation image contrast resulting from the selective absorption of
        the primary beam by the patient is called:
        A. Subjective contrast
        B. Objective contrast
        C. Photographic contrast
        D. All of the above   Ref. 13 - p. 152

961.   That part of the radiation contrast than can be seen recorded on the film
is called:
A.   Subjective contrast
B.   Objective contrast
C.   Photographic contrast
D.   All of the above   Ref. 13 - p. 152

962.   Objective contrast cannot be increased in radiography by:
A.   Opaque media
B.   Lower Kv
C.   Correct development
D.   Spotlighting   Ref. 13 - p. 187

963.   Fluoroscopic screens emit a yellowish greenish light because:
A.   It is more pleasant to the viewer
B.   It gives more contrast
C.   Photographic film is more sensitive to it
D.   It is the light to which the human eye is most sensitive
Ref. 13 - p. 189

964.   The limit of permissible tube current in fluoroscopy is:
A.   10-25 mA
B.   100-200 mA
C.   2-4 mA
D.   Any of the above   Ref. 13 - p. 191

965.   In most countries the regulations for lead equivalency of the lead glass
window of a fluoroscopic screen is:
A.   1.5mm up to 100 Kv
B.   3mm up to 100 Kv
C.   2.5mm between 100 Kv to 150 Kv
D.   A and C   Ref. 13 - p. 191

966.   Compared to a normal fluoroscopic screen, the graininess of the viewing
screen of an image intensifier is:
A.   Greater
B.   Smaller
C.   The same
D.   Non-existent   Ref. 13 - p. 193

967.   The opacity of a film is defined as:
A.   The density of the metallic silver
B.   The ratio of amount of light incident on the film to the amount
transmitted by the film
C.   The density of silver bromide crystals per unit area
D.   The absorption percentage of the part examined
Ref. 13 - p. 195

968.   The logarith to base 10 of the opacity defines:
A.   Translucency
B.   Contrast
C.   Density
D.   Visibility of detail   Ref. 13 - p. 195

969.   The base of the film:
A.   Does not add any density to the image
B.   Contributes a density from 0.06 to 0.2
C.   Adds appreciable density to the film
D.   None of the above   Ref. 13 - p. 196

970. One of the following is incorrect:
   A. The sensitivity of the film has no influence on fog density
   B. Fog density is increased by overdevelopment
   C. Temperature at which films are stored influences fog density
   D. Incorrect darkroom lighting leads to more fog
   Ref. 13 - p. 197

971. The H and D characteristic curve:
   A. Demonstrates selective absorption
   B. Indicates heat dissipation by the tube
   C. Gives the relation between exposure and the resulting density
   D. Shows the spectrum of an heterogeneous beam
   Ref. 13 - p. 197

972. The tangent of the angle which the straight portion of the characteristic curve makes with the abscissa is called:
   A. Gamma
   B. Delta
   C. Gradation
   D. A and C
   Ref. 13 - p. 198

973. Gradation can be improved by:
   A. Overdeveloping
   B. Overexposing
   C. Use of double coated films
   D. Use of single coated films
   Ref. 13 - p. 199

974. Compared to slow films, fast films have:
   A. A steeper gradation
   B. A less steep gradation
   C. The same gradation
   D. A higher gamma
   Ref. 13 - p. 202

975. The difference in density between the extreme points in the useful area of a characteristic curve:
   A. Is called the density range
   B. Is directly related to the exposure range
   C. Shows the exposure latitude
   D. All of the above
   Ref. 13 - p. 203

976. The object contrast range must be _____ than the exposure range if the object is to be adequately demonstrated:
   A. Greater
   B. Smaller
   C. Equal
   D. Is of no importance
   Ref. 13 - p. 203

977. In films with a steep gradation:
   A. The correct exposure range is large
   B. Exposure must be chosen more accurately
   C. Lowering Kv, reduces object contrast
   D. Record long scale ranges at low Kv
   Ref. 13 - p. 204

978. Films with steep gradation are nearly always used and to decrease the object contrast range one may:
   A. Increase Kv
   B. Use a compensating filter
   C. Use graduated intensifying screens
   D. All of the above
   Ref. 13 - p. 205

979.   If for 40" FFD, the exposure requires 50 mAs, what would be the mAs
        at 36":
        A. 25 mAs
        B. 40.5 mAs
        C. 45 mAs
        D. 30 mAs   Ref. 13 - p. 251

980.   An exposure without grid requires 20 mAs; what would be the mAs
        when using a grid having a grid factor of 2.5:
        A. 75 mAs
        B. 25 mAs
        C. 50 mAs
        D. 40 mAs   Ref. 13 - p. 252

981.   If an exposure of 100 mAs is necessary without screens, then the exposure
        with screens having an intensification factor of 20 will be:
        A. 10 mAs
        B. 7.5 mAs
        C. 5 mAs
        D. 2 mAs   Ref. 13 - p. 259

982.   A switching error of 2 mAs in 5 mAs means an error of:
        A. 40%
        B. 60%
        C. 20%
        D. 35%   Ref. 13 - p. 261

983.   Each centimeter increase in object thickness requires:
        A. An increase of 5% in Kv
        B. An increase of 25% of the mAs
        C. An increase of 25% in the exposure value
        D. All of the above   Ref. 13 - p. 264

984.   If one wishes to reproduce with a 4 valve (single phase), the exposure
        valid for a three phase unit then one must choose:
        A. 10% higher Kv and 20% higher mAs
        B. 10% higher Kv
        C. 20% higher mAs
        D. 10% lower Kv and 25% higher mAs   Ref. 13 - p. 265

985.   Exposure data can be influenced by:
        A. Type of film
        B. Anode material
        C. Filtration
        D. All of the above   Ref. 13 - p. 265

986.   If a radiograph is clearly underexposed, the exposure should be:
        A. Slightly longer
        B. Doubled
        C. Increased by 25%
        D. Increased by 75%   Ref. 13 - p. 264

987.   The inherent filtration of the tube:
        A. Affects the emergent radiation both quantitatively and qualitatively
        B. Does not affect exposure
        C. Only affects the quality of radiation
        D. Only affects the amount of radiation   Ref. 13 - p. 266

988. One of the following does not contribute to extra filtration of the X-ray beam:
   A. Top of table
   B. Covering on the table
   C. Inherent filtration
   D. Cassette front   Ref. 13 - p. 266

989. Identical exposures for two X-ray tubes:
   A. Will always produce the same densities
   B. Will be different due to differences in space charge value
   C. Will be different due to different anode size
   D. None of the above   Ref. 13 - p. 267

990. The ionization chamber phototimer:
   A. Uses ionization chamber between the patient and the film
   B. Terminates the exposure by capacitor discharge
   C. The capacitor is preset to discharge when a charge is reached
   D. All of the above   Ref. 13 - p. 294

991. The part of the object called "dominant" by Franke is:
   A. The most important part of the radiograph
   B. The part seen by the "pick up" of the phototimer
   C. The part determining the density of the film
   D. All of the above   Ref. 13 - p. 294

992. One of the following is not true when using phototimers:
   A. Processing technique must be strictly standardized
   B. Any film can be used
   C. Intensifying screens must be standard
   D. Same type of cassettes should be used
                                               Ref. 13 - p. 295

993. A draw back in using the automatic exposure device is that:
   A. The Kv for each category must be based on the thickest object likely
      to be encountered
   B. The mAs does not follow the same rule
   C. One has to use high mA
   D. Kv, mA and time for each category must be based upon the thickest
      object likely to be encountered   Ref. 13 - p. 297

994. The preset exposure in a phototimer is relatively long: in cases where
   exposure time is shorter:
   A. The film is always perfect
   B. The film is always denser than normal
   C. The film is always lighter than normal
   D. The tube is not optimally loaded   Ref. 13 - p. 298

995. The falling load method consists of:
   A. The anode reaches almost immediately its maximum permissible
      temperature
   B. After exposure is initiated the current falls gradually
   C. Keeping the energy supplied to the anode in equilibrium with the
      energy given up by it
   D. All of the above   Ref. 13 - p. 298

996. If there is less difference between the attenuation coefficient of the
   various absorbent substances in the body:
   A. Higher tensions are used
   B. The radiation contrast is decreased
   C. Scatter has increased
   D. All of the above   Ref. 13 - p. 300

(T)RUE OR (F)ALSE

997.   Using high Kv technique; patient dose is lower:
                                Ref. 13 - p. 301

998.   Using high Kv technique a smaller focus cannot be used.
                                 Ref. 13 - p. 302

999.   Using high Kv technique exposure time may be shortened.
                                 Ref. 13 - p. 303

1000.   Techniques in the 30 Kv - 50 Kv range are never used.
                                 Ref. 13 - p. 305

1001.   The pulse technique is used for cine fluorography.
                                 Ref. 13 - p. 305

1002.   Any X-ray tube may be used for pulse technique.
                                 Ref. 13 - p. 306

1003.   With high temperature development the developing time is prolonged.
                                 Ref. 10 - p. 54

1004.   Direct exposure films can be used with intensifying screens.
                                 Ref. 10 - p. 67

1005.   Intensifying screens should never be cleaned with common cleaning fluids.
                                 Ref. 10 - p. 74

FOR EACH OF THE FOLLOWING MULTIPLE CHOICE QUESTIONS
SELECT THE ONE MOST APPROPRIATE ANSWER:

1006.   Fundamental units employed in physics are:
        A. Length, mass and time
        B. Area, volume and velocity
        C. Energy, acceleration and gravity
        D. All of the above   Ref. 6 - p. 27

1007.   Unit of length in metric system is:
        A. Meter
        B. Liter
        C. Kilogram
        D. Kilometer   Ref. 6 - p. 27

1008.   The formula for conversion of degrees fahrenheit to degrees
        centigrade is:
        A. 9/5 C + 32
        B. 5/9 (F - 32)
        C. 9/5 F - 32
        D. 9/5 F + 32   Ref. 6 - p. 30

1009.   The product of the distance an object is moved and the force applied
        to move it equals:
        A. Energy
        B. Mass
        C. Velocity
        D. Work   Ref. 6 - p. 31

1010.   The law of conservation of energy states that the total amount of
        energy in the universe is:
        A. Constant
        B. Fluctuating
        C. Diminishing
        D. Increasing   Ref. 6 - p. 34

1011.   The unit of mass in the metric system is the:
        A. Meter
        B. Liter
        C. Pound
        D. Kilogram   Ref. 6 - p. 28

1012.   The mass per unit volume of a substance is the:
        A. Area
        B. Volume
        C. Velocity
        D. Density   Ref. 6 - p. 29

1013.   Specific gravity is the ratio of the density of any substance to the
        density of:
        A. Water
        B. Air
        C. Oxygen
        D. Lead   Ref. 6 - p. 29

1014.   The temperature of a solution is 68 degrees F. What is the temperature
         in centigrade?:
         A. 65 degrees C
         B. 20 degrees C
         C. 55 degrees C
         D. 80 degrees C   Ref. 6 - p. 30

1015.   To set a resting body in motion, one must apply:
         A. Potential energy
         B. Work
         C. Force
         D. Gravity   Ref. 6 - p. 31

1016.   The energy of motion is called:
         A. Kinetic
         B. Potential
         C. Stored
         D. Inherent   Ref. 6 - p. 31

1017.   The smallest subdivision of a substance having the physical properties
         of that substance is a(an):
         A. Molecule
         B. Atom
         C. Electron
         D. Element   Ref. 6 - p. 37

1018.   The smallest particle of an element having its characteristic
         properties, that can chemically combine with atoms of another element
         is a(an):
         A. Atom
         B. Compound
         C. Mixture
         D. Molecule   Ref. 6 - p. 37

1019.   A substance which cannot be reduced to a simpler substance by
         ordinary chemical means is called a(an):
         A. Molecule
         B. Element
         C. Compound
         D. Mixture   Ref. 6 - p. 37

1020.   The particles which revolve in orbits around the nucleus of an atom
         are the:
         A. Neutrinos
         B. Neutrons
         C. Protons
         D. Electrons   Ref. 6 - p. 39

1021.   The particles, within the nucleus, that contribute a positive charge
         are the:
         A. Neutrons
         B. Protons
         C. Positrons
         D. Electrons   Ref. 6 - p. 41

1022.   The particle within the nucleus of an atom that has no charge is the:
         A. Shell
         B. Electron
         C. Proton
         D. Neutron   Ref. 6 - p. 41

1023.   The atomic number of an atom is equal to the number of its:
    A.   Electrons
    B.   Protons
    C.   Neutrons
    D.   Positrons   Ref. 6 - p. 42

1024.   The mass number of an atom is equal to the number of its:
    A.   Neutrons
    B.   Protons
    C.   Neutrons and protons
    D.   Neutrons and electrons   Ref. 6 - p. 42

1025.   Isotopes are atoms of the same element that differ from one another
    in the number of:
    A.   Neutrons
    B.   Protons
    C.   Neutrons and protons
    D.   Orbital shells   Ref. 6 - p. 42

1026.   The atomic weight refers to the mass of an atom relative to the mass
    of an atom of:
    A.   Hydrogen
    B.   Oxygen
    C.   Helium
    D.   Water   Ref. 6 - p. 43

1027.   All of the elements can be arranged in an orderly series on the basis
    of atomic weight or atomic number. This series is called the:
    A.   Half value table
    B.   Periodic table
    C.   Isodose curve
    D.   Absorption coeficient   Ref. 6 - p. 44

1028.   The simplest element, consisting of one proton and one electron is:
    A.   Hydrogen
    B.   Oxygen
    C.   Carbon
    D.   Helium   Ref. 6 - p. 44

1029.   Charged atoms are known as:
    A.   Electrons
    B.   Electricity
    C.   Ions
    D.   Electrolysis   Ref. 6 - p. 51

1030.   The process of changing a neutral atom to an electrically charged
    particle is known as:
    A.   Ionization
    B.   Neutralization
    C.   Electrification
    D.   All of the above   Ref. 6 - p. 51

1031.   The negative pole of a battery is known as the:
    A.   Anode
    B.   Cathode
    C.   Anion
    D.   Cation   Ref. 6 - p. 52

1032. The ejection of electrons when a metal is heated to a glow is called:
       A. Electrolysis
       B. Chemical ionization
       C. Thermionic emission
       D. Metal fatigue   Ref. 6 - p. 52

1033. Materials, such as metals, which allow a free flow of electrons are
       called:
       A. Transducers
       B. Transformers
       C. Insulators
       D. Conductors   Ref. 6 - p. 55

1034. If an uncharged metal is brought into the electrostatic field of a charged
       object, there will be an electron shift in the uncharged metal. This
       is called:
       A. Induction
       B. Friction
       C. Contact
       D. Magnification   Ref. 6 - p. 55

1035. Any electrical charge can be neutralized if it is conducted to:
       A. A magnet
       B. Ground
       C. The positive pole
       D. The negative pole   Ref. 6 - p. 57

1036. Two bodies of like electrical charge will _____ each other:
       A. Magnify
       B. Neutralize
       C. Repel
       D. Attract   Ref. 6 - p. 57

1037. Electric charges are present only in what portion of conductor?:
       A. Internal surface
       B. External surface
       C. Center
       D. Throughout the conductor   Ref. 6 - p. 59

1038. If electrons jump from a more negative to a less negative body, what is
       produced?:
       A. Valence
       B. Neutron
       C. Field
       D. Spark   Ref. 6 - p. 62

1039. An electric current in a solid conductor consists of a flow of:
       A. Waves
       B. Neutrons
       C. Protons
       D. Electrons   Ref. 6 - p. 65

1040. What is the name of the path over which electric current flows?:
       A. Potential
       B. Transmitter
       C. Circuit
       D. Resistance   Ref. 6 - p. 65

1041.   The maximum difference of potential between the terminals of an
        electrical generation is called:
        A. Amperage
        B. Wattage
        C. Electromotive force
        D. Power   Ref. 6 - p. 68

1042.   The amount of electricity flowing per second is called the:
        A. Current
        B. Potential difference
        C. Resistance
        D. Power   Ref. 6 - p. 68

1043.   The unit of potential difference is called the:
        A. Coulomb
        B. Watt
        C. Ampere
        D. Volt   Ref. 6 - p. 68

1044.   The potential difference which will cause a current of one ampere to
        flow in a circuit of one OHM resistance is known as the:
        A. Volt
        B. Ampere
        C. Watt
        D. Joule   Ref. 6 - p. 68

1045.   That property which opposes the flow of electric current is called:
        A. Magnetism
        B. Resistance
        C. Potential
        D. Reduction   Ref. 6 - p. 68

1046.   The measuring device used to determine the quantity of electricity
        flowing per second:
        A. Filament control
        B. Flow meter
        C. Voltmeter
        D. Ammeter   Ref. 6 - p. 72

1047.   An electric circuit whose parts are arranged in a row is said to be of
        what type?:
        A. Series
        B. Parallel
        C. Complex
        D. None of the above   Ref. 6 - p. 74

1048.   I = V is an expression of what law of electricity?:
            R
        A. Joule's
        B. Einstein's
        C. Circuit
        D. Ohm's   Ref. 6 - p. 75

1049.   If 2400 volts are applied across a resistance of 120 OHMs, the current
        flowing will be:
        A. 50 amperes
        B. 20 amperes
        C. 20 milliamperes
        D. 500 amperes   Ref. 6 - p. 75

1050. The quantity of electricity stored in a capacitor is determined by:
A. Area of capacitor plates
B. Distance between plates
C. Material used as insulator between plates
D. All of the above   Ref. 6 - p. 81

1051. The power of a steady direct current of 50 amperes and 10 volts is:
A. 500 ohms
B. 500 watts
C. 5 watts
D. 5 joules   Ref. 6 - p. 82

1052. The power loss in the form of heat is proportional to the:
A. Amperage
B. Square of the voltage
C. Square of the amperage
D. Voltage   Ref. 6 - p. 82

1053. An example of a natural magnet is which one of the following?:
A. The earth
B. Lead
C. Hard steel
D. Electromagnet   Ref. 6 - p. 84

1054. If the distance between two magnetic poles is doubled the magnetic force between them is:
A. Decreased by 1/4
B. Decreased by 1/2
C. Doubled
D. Tripled   Ref. 6 - p. 86

1055. A _____ always exists around a conductor in which a current is flowing:
A. Resistance
B. Negative charge
C. Magnetic field
D. Positive charge   Ref. 6 - p. 94

1056. Voltage can be induced in a wire by:
A. Movement of the wire through a magnetic field
B. Movement of a magnetic field across a wire
C. Varying the strength of a magnetic field while a wire is in it
D. All of the above   Ref. 6 - pp. 98, 99

1057. A device which converts mechanical energy into electrical energy using the principle of electromagnetic induction is which of the following?:
A. Generator
B. Dynamo
C. Turbine
D. All of the above   Ref. 6 - p. 103

1058. A current which repeatedly reverses its direction is called:
A. Direct
B. Alternating
C. Induced
D. Multiphase   Ref. 6 - p. 106

1059. When a wire carrying electric current is put in a magnetic field, there
       is force exerted on the wire to:
       A. Keep it stationary
       B. Move it out of the magnetic field
       C. Move it toward the north magnetic pole
       D. Move it toward the south magnetic pole
                                                    Ref. 6 - p. 113

1060. A device that converts electrical energy to mechanical energy is:
       A. Solenoid
       B. Generator
       C. Electric motor
       D. Capacitor   Ref. 6 - p. 113

1061. The electrical device which increases the incoming alternating current
       from a low voltage to a high voltage is a(an):
       A. Capacitor
       B. Electrodynamometer
       C. Transformer
       D. Coil   Ref. 6 - p. 120

1062. The X-ray generator puts out a higher voltage than that applied to it
       by a process known as:
       A. Induction
       B. Discharge
       C. Current flow
       D. Generation   Ref. 6 - p. 121

1063. Power loss in a transformer due to heat production within the core
       is called:
       A. Copper loss
       B. Core loss
       C. Eddy current loss
       D. Hysteresis loss   Ref. 6 - p. 126

1064. A device used to regulate voltage and amperage in the filament circuit of
       an X-ray tube is:
       A. Choke coil
       B. Rheostat
       C. Autotransformer
       D. None of the above   Ref. 6 - p. 132

1065. The term rectification refers to the:
       A. Changing of an alternating into a direct current
       B. Increase of voltage in a circuit
       C. Control of transformer loss
       D. None of the above   Ref. 6 - p. 136

1066. That portion of the X-ray tube consisting of a metal filament that emits
       electrons when heated is the:
       A. Housing
       B. Cathode
       C. Anode
       D. Target   Ref. 6 - p. 136

1067. The high voltage current passes through an X-ray tube in which
       direction?:
       A. Anode to cathode
       B. Cathode to anode
       C. Both directions
       D. Neither direction   Ref. 6 - p. 137

1068. In a self-rectified circuit, the _____ is applied directly to the
       X-ray tube terminals:
       A. Heat
       B. Resistance
       C. Current
       D. High voltage   Ref. 6 - p. 138

1069. Self-rectification is used in which type of X-ray apparatus?:
       A. High voltage
       B. Portable unit
       C. Supervoltage
       D. None of the above   Ref. 6 - p. 138

1070. Self-rectification fails if the _____ is heated so that electron
       emission occurs:
       A. Anode
       B. Cathode
       C. Transformer
       D. All of the above   Ref. 6 - p. 138

1071. A greater load can be applied to an X-ray tube if which type of
       rectification is used?:
       A. Self
       B. Self-half wave
       C. Half wave
       D. Full wave   Ref. 6 - p. 146

1072. Roentgen rays travel at the same constant speed as:
       A. Sound
       B. Air
       C. Light
       D. All of the above   Ref. 6 - p. 149

1073. Roentgen rays are one type of _____ wave:
       A. Sine
       B. Sound
       C. Electromagnetic
       D. Heat   Ref. 6 - p. 149

1074. The speed that X-rays travel equals their wavelength multiplied by:
       A. Frequency of vibration
       B. Absorption coefficient
       C. Inverse voltage
       D. Alternating current   Ref. 6 - p. 150

1075. The useful range of X-ray wavelengths is about:
       A. 0.001-0.005 Angstrom
       B. 0.1-0.5 Angstrom
       C. 1-5 Angstrom
       D. 10-50 Angstrom   Ref. 6 - p. 150

1076. The longest type of electromagnetic wave is the:
       A. Ultraviolet
       B. Infra-red
       C. X-ray
       D. Radio   Ref. 6 - p. 151

1077. Whenever a stream of fast moving electrons are suddenly reduced in speed, what kind of rays are produced?:
A. Light
B. Sound
C. X-rays
D. Radio   Ref. 6 - p. 151

1078. X-rays are physically identical with:
A. Gamma rays
B. Beta rays
C. Alpha rays
D. Delta rays   Ref. 6 - p. 151

1079. The following substance must be removed from an X-ray tube before it can be used:
A. Air
B. Carbon
C. Lead
D. All of the above   Ref. 6 - p. 152

1080. About _____ percent of the kinetic energy of the electron stream between cathode and anode is converted to X-rays:
A. 99.8
B. 20
C. 2
D. 0.2   Ref. 6 - p. 155

1081. When an electron is ejected from an orbit and another electron fills its space, X-ray energy is given off. What is this energy called?:
A. Dissipation
B. Static
C. Thermionic emission
D. Characteristic radiation   Ref. 6 - p. 156

1082. Target metal used in an X-ray tube must have which property?:
A. High melting point
B. High atomic number
C. Both of the above
D. Neither of the above   Ref. 6 - p. 157

1083. The following properties apply to X-rays:
A. Highly penetrating
B. Electrically neutral
C. Occur in wide range of wave lengths
D. All of the above   Ref. 6 - p. 157

1084. The following are properties characteristic of X-rays, except:
A. Emerge in straight lines from tube
B. Cannot be focussed by a lens
C. Produce biological changes
D. Never affect photographic film   Ref. 6 - p. 158

1085. The radiation exposure per unit time at a given location is known as the:
A. Exposure dose
B. Exposure rate
C. Exposure factor
D. Exposure ratio   Ref. 6 - p. 159

1086. The total radiation exposure is equal to the exposure time multiplied by the:
A. Exposure dose
B. Background count
C. Exposure rate
D. Uptake
Ref. 6 - p. 159

1087. The instrument used to calibrate the roentgen output of a radiation therapy unit is:
A. Standard free air ionization chamber
B. Victoreen R-meter
C. Geiger-Mueller tube
D. Oscilloscope
Ref. 6 - p. 160

1088. Exposure rate may be changed by varying:
A. Milliamperage
B. Kilovoltage
C. Filtration
D. Any of the above
Ref. 6 - p. 160

1089. The minimum wavelength of an X-ray beam depends only on the:
A. Inverse square law
B. Tube rating
C. Peak kilovoltage
D. Amperage
Ref. 6 - p. 167

1090. The term "half value layer" describes which characteristic of an X-ray beam?:
A. Direction
B. Quality
C. Speed
D. Quantity
Ref. 6 - p. 168

1091. If the exposure rate from an X-ray unit is initially 100 r/minute, and a filter of 2 mm Al is added and is found to decrease this rate to 50 r/minute, then the half value layer would be:
A. 4 mm Al
B. 2 mm Al
C. 1/2
D. 50 r/minute
Ref. 6 - p. 169

1092. "Soft" X-rays have a relatively:
A. Lower frequency and penetrating power
B. Longer wavelength
C. Greater absorption in the skin
D. All of the above
Ref. 6 - p. 172

1093. "Soft" X-rays may be produced by:
A. Lower kilovoltage
B. Lighter filtration
C. Targets of low atomic number
D. All of the above
Ref. 6 - p. 172

1094. "Hard" X-rays may be produced by:
A. Lower kilovoltage
B.
C. Targets of high atomic number
D. All of the above
Ref. 6 - p. 172

1095.   X-rays that have changed direction following collision with atoms
        are called:
        A.   Primary beam
        B.   Scattered radiation
        C.   Secondary radiation
        D.   Absorbed radiation   Ref. 6 - p. 172

1096.   The type of X-ray interaction in which all of the energy of the incident
        photon is expended in dislodging a bound electron is called:
        A.   Photoelectric effect
        B.   Compton effect
        C.   Pair production
        D.   Coherent scattering   Ref. 6 - p. 175

1097.   When a filter is used to "harden" an X-ray beam, it does so mainly by:
        A.   Photoelectric absorption
        B.   Compton effect
        C.   Pair production
        D.   Unmodified scatter   Ref. 6 - p. 180

1098.   Photoelectric absorption is about 6 times greater in bone than in soft
        tissue in the diagnostic kilovoltage range. This is responsible for the
        _____ between these 2 types of tissues:
        A.   Energy levels
        B.   Tone
        C.   Resolution
        D.   Radiographic contrast   Ref. 6 - p. 181

1099.   In the usual diagnostic kilovoltage range, which type of interaction
        between radiation and matter predominates?:
        A.   Compton effect
        B.   Photoelectric absorption
        C.   Pair production
        D.   Unmodified scatter   Ref. 6 - p. 181

1100.   The scattered radiation during a diagnostic X-ray examination consists
        mainly of the:
        A.   Compton effect
        B.   Photoelectric absorption
        C.   Pair production
        D.   All of the above   Ref. 6 - p. 179

1101.   Tissue changes caused by X-rays are thought to be due to transfer of
        energy to tissue atoms by the process of:
        A.   Absorption
        B.   Excitation
        C.   Ionization
        D.   All of the above   Ref. 6 - p. 182

1102.   The unit used to specify radiation exposure is the:
        A.   Rad
        B.   Roentgen
        C.   Milamperage
        D.   Kilovoltage   Ref. 6 - p. 183

1103.   The surface radiation exposure rate is _____ the in air-radiation
        exposure rate:
        A.   Less than
        B.   Greater than
        C.   Equal to   Ref. 6 - p. 185

1104.   If the area of the field and thickness of the irradiated part are increased
         the backscatter factor is then:
         A. Increased
         B. Decreased
         C. Unchanged   Ref. 6 - p. 186

1105.   Backscatter becomes negligible at what voltage range?:
         A. 20-50 Kv
         B. 50-150 Kv
         C. 250-300 Kv
         D. Above 1 Mv   Ref. 6 - p. 186

1106.   With cobalt 60 beam therapy, the maximum exposure rate is found
         where in the body?:
         A. 5 mm above the skin surface
         B. At the skin surface
         C. 5 mm below skin surface
         D. Midpoint of body   Ref. 6 - p. 187

1107.   The law which states that the "exposure rate is inversely proportional
         to the square of the distance between the radiation source and the point
         of interest, " is called:
         A. Ohm's law
         B. Inverse square law
         C. Law of conservation of energy
         D. Proportionality law   Ref. 6 - p. 187

1108.   The amount of radiation delivered to a tumor is best expressed in
         terms of:
         A. Air dose
         B. Skin dose
         C. Percent depth dose
         D. Erythema dose   Ref. 6 - p. 188

1109.   As the treatment field area is increased, what happens to the %
         depth dose?:
         A. Decreases
         B. Increases
         C. Remains unchanged   Ref. 6 - p. 191

1110.   As the radiation source to surface distance (SSD) is increased, what
         happens to the % depth dose?:
         A. Decreases
         B. Increases
         C. Remains unchanged   Ref. 6 - p. 191

1111.   A filter placed in the path of an X-ray beam produces what effect?:
         A. Decreases the hardness of the beam
         B. Alters the minimum wavelength
         C. Increases the hardness of the beam
         D. Removes a greater proportion of high energy waves
                                                 Ref. 6 - p. 192

1112.   When a filter is added to an X-ray beam, which of the following factors
        is always reduced?:
        A. Exposure rate
        B. Hardness of the beam
        C. % depth dose
        D. Penetrating power of the beam   Ref. 6 - p. 194

1113.   The primary filter used in the 150-400 Kv range is composed of:
        A. Lead
        B. Aluminum
        C. Tin
        D. Copper   Ref. 6 - p. 194

1114.   The maximum wave length in a heterogenous X-ray beam is determined by:
        A. Peak kilovoltage
        B. Type and amount of added filtration
        C. Inherent filtration of the tube
        D. Milliamperage   Ref. 6 - p. 194

1115.   In an X-ray tube, electrons are produced when the _____ is heated:
        A. Glass envelope
        B. Housing
        C. Anode
        D. Cathode   Ref. 6 - p. 198

1116.   In an X-ray tube, if a kilovoltage is applied, the electrons are driven
        to the:
        A. Glass envelope
        B. Filament
        C. Anode
        D. Cathode   Ref. 6 - p. 198

1117.   When electrons strike the target of an X-ray tube, most of their
        energy is converted to:
        A. X-rays
        B. Heat
        C. Light
        D. Ultraviolet rays   Ref. 6 - p. 198

1118.   The filament current in an X-ray tube is used to provide:
        A. A source of electrons
        B. Potential to drive electrons to the anode
        C. A source of heat dissipation
        D. Prevention of sparkover   Ref. 6 - p. 200

1119.   The filament current in an X-ray tube operates at about what voltage?:
        A. 10 volts
        B. 50 volts
        C. 100 volts
        D. 250 volts   Ref. 6 - p. 200

1120.   Diagnostic X-ray tubes containing two filaments are known as:
        A. Stationary anode tubes
        B. Valve tubes
        C. Rotating anode tubes
        D. Double focus tubes   Ref. 6 - p. 200

1121. The acutal target of an X-ray tube is composed of what substance?:
A. Lead
B. Tungsten
C. Copper
D. Tin   Ref. 6 - p. 202

1122. The part of the X-ray tube target actually bombarded by the electrons is the:
A. Focus
B. Filament
C. Copper block
D. Housing   Ref. 6 - p. 202

1123. The type of X-ray tube in which the target continually "turns a new face" to the electron beam during the exposure is called a _____ tube:
A. Double focus
B. Valve
C. Rotating anode
D. Stationary anode   Ref. 6 - p. 204

1124. If no voltage is applied across an X-ray tube and electrons remain as a "cloud" around the hot filament, the resultant effect is called:
A. Spark gap
B. Noise
C. Gassing
D. Space charge   Ref. 6 - p. 206

1125. When the tube current (Ma) is plotted against the applied tube voltage (Kv), a _____ is obtained:
A. Characteristic curve
B. Isodose curve
C. Time-temperature curve
D. Absorption curve   Ref. 6 - p. 207

1126. The point at which further increase in kilovoltage causes no additional increase in milliamperage is known as:
A. End point
B. Space change point
C. Absorption point
D. Saturation point   Ref. 6 - p. 207

1127. When an X-ray tube is operating at the maximum tube current for a given filament current, it is said to be operating on:
A. Fixed potential
B. Alternating current
C. Saturation current
D. Effective current   Ref. 6 - p. 207

1128. Most X-ray tube failures are due to:
A. Pitting in the target
B. Burning out the target
C. Gas
D. Burning out the filament   Ref. 6 - p. 208

1129. A _____ circuit is usually added to the filament circuit to limit the filament current until the X-ray exposure switch is closed:
A. Booster
B. Full wave rectified
C. Half wave rectified
D. Parallel   Ref. 6 - p. 208

1130.   An increase in kilovoltage produces a _____ load on the X-ray tube
        as compared with a similar increase in milliampere-seconds:
        A. Greater
        B. Smaller
        C. Similar   Ref. 6 - p. 209

1131.   A tube rating chart is used to determine:
        A. Safe limits of tube operation
        B. Output of the tube
        C. Tube current as a function of applied voltage
        D. Tube size   Ref. 6 - p. 208

1132.   The maximum safe exposure for an X-ray tube is _____ with full
        wave rectification than with self-rectification:
        A. Less
        B. Greater
        C. The same   Ref. 6 - p. 210

1133.   The maximum safe exposure for an X-ray tube is _____ with a
        small focus than with a large one:
        A. Less
        B. Greater
        C. The same   Ref. 6 - p. 210

1134.   The ability of the anode to anode and dissipate heat is measured in:
        A. Thermal units
        B. Heat units
        C. Fahrenheit
        D. Centigrade   Ref. 6 - p. 210

1135.   The measurement of the anode's ability to store and dissipate that can
        be obtained by multiplying which of the following factors?:
        A. Kv X Ma
        B. Kv x Sec.
        C. Sex. x Ma
        D. Ma x Kv x Sec.   Ref. 6 - p. 210

1136.   The heat storage capacity of the anode of various diagnostic X-ray tubes
        ranges from about:
        A. 25,000-50,000 H.U.
        B. 75,000-100,000 H.U.
        C. 100,000-250,000 H.U.
        D. 500,000-750,000 H.U.   Ref. 6 - p. 211

1137.   The cooling characteristics of an X-ray tube are referred to as the:
        A. Heat dissipation rate
        B. Heat storage capacity
        C. Cold or hot tube
        D. Melting point   .Ref. 6 - p. 211

1138.   With minor overloading of the X-ray tube, little craters form on the
        target which is known as:
        A. Evaporation of metal
        B. Pitting
        C. Line focus
        D. Erosion   Ref. 6 - p. 213

1139.   An X-ray tube is usually immersed in oil for the purpose of:
        A.   Insulation
        B.   Heat retention
        C.   Protection
        D.   Cooling   Ref. 6 - p. 213

1140.   Superficial X-ray therapy units operate at what potential?:
        A.   20-30 Kv
        B.   50-120 Kv
        C.   130-150 Kv
        D.   200-250 Kv   Ref. 6 - p. 213

1141.   Orthovoltage X-ray therapy units operate at what potential?:
        A.   20-30 Kv
        B.   50-120 Kv
        C.   130-150 Kv
        D.   200-250 Kv   Ref. 6 - p. 214

1142.   Supervoltage therapy units operate at which kilovoltage range?:
        A.   50-100 Kv
        B.   150-250 Kv
        C.   300-500 Kv
        D.   Above 1 Mv   Ref. 6 - p. 214

1143.   The difference between a diagnostic and a therapy X-ray tube is that
        in the therapy tube there is always:
        A.   A rotating anode
        B.   A smaller focal spot
        C.   A larger filament and focal spot and also a stationary anode
        D   A smaller filament   Ref. 6 - p. 214

1144.   In some radiation therapy units, both the tube and transformer are
        mounted in the same casing, suspended above the patient; this is
        called a(an):
        A.   Water-cooled unit
        B.   Self-contained unit
        C.   Oil-cooled unit
        D.   Portable unit   Ref. 6 - p. 214

1145.   The difference between a valve tube and an X-ray tube is that the
        valve tube:
        A.   Operates below saturation
        B.   Is a thermionic diode tube
        C.   Has a shorter and smaller filament
        D.   Has two circuits   Ref. 6 - p. 215

1146.   A valve tube operates at _____ milliamperage as the X-ray
        tube to which it is connected in a series circuit:
        A.   The same
        B.   Greater
        C.   Lesser   Ref. 6 - p. 216

1147.   In a four-valve tube rectified circuit the peak milliamperage is
        approximately _____ times the average milliamperage:
        A.   1-1/2
        B.   2
        C.   2-1/2
        D.   3   Ref. 6 - p. 225

1148.   Assuming that techniques are used to produce identical density and
         contrast, radiation exposure to the patient with three phase equipment
         is _____ with standard equipment:
         A. The same as
         B. Greater than
         C. Less than   Ref. 6 - p. 245

1149.   The voltage supplied to the equipment in a radiology department
         is called:
         A. Effective voltage
         B. Peak voltage
         C. Line voltage
         D. Filament voltage   Ref. 6 - p. 228

1150.   The incoming current to a radiology department is of what type?:
         A. Direct
         B. Alternating
         C. Three phase
         D. Rectified   Ref. 6 - p. 229

1151.   That part of the circuit connected to the primary transformer coil is
         called the _____ circuit, while that part connected to the secondary
         coil is called the _____ circuit:
         A. Low voltage, high voltage
         B. High voltage, low voltage
         C. Secondary, primary
         D. Incoming, outgoing   Ref. 6 - p. 229

1152.   When the X-ray circuit is overloaded, the circuit is broken and the
         equipment thereby protected. This is done by inserting _____ into
         the circuit:
         A. Transformers
         B. Coils
         C. Fuses
         D. Rheostats   Ref. 6 - p. 229

1153.   Which type of timer operates in the range of 1/20 sec. to 20 seconds
         exposure?:
         A. Impulse
         B. Electronic timer
         C. Manual timer
         D. Synchronous timer   Ref. 6 - p. 232

1154.   Which type of timer operates in the range of 1/20 sec. to 1/4 sec.
         exposure?:
         A. Impulse timer
         B. Electronic timer
         C. Manual timer
         D. Synchronous timer   Ref. 6 - p. 233

1155.   The automatic exposure device used most frequently for spot filming
         during fluoroscopy is the:
         A. Image intensifier
         B. Phototimer
         C. Photofluorography
         D. Television viewer   Ref. 6 - p. 233

1156.   Another device used to prevent overloading of the X-ray circuit is the:
        A. Timer
        B. Main switch
        C. Rheostat
        D. Circuit breaker   Ref. 6 - p. 234

1157.   The X-ray tube current is measured by the:
        A. Filament stabilizer
        B. Phototimer
        C. Milliammeter
        D. Voltmeter   Ref. 6 - p. 237

1158.   The following system is used to change the alternating current supplied
        by the transformer into direct current:
        A. Rectification
        B. Three phase
        C. Step up
        D. Step down   Ref. 6 - p. 238

1159.   In modern shock-proof X-ray equipment, the following component is
        grounded:
        A. Mid-point of primary coil
        B. Mid-point of secondary coil
        C. Cables
        D. Autotransformer   Ref. 6 - p. 238

1160.   The controls and meters of X-ray equipment are mounted in the:
        A. Tube housing
        B. Generator
        C. Lead screen
        D. Control panel   Ref. 6 - p. 239

1161.   The current which heats the X-ray tube filament is registered by the:
        A. Milliammeter
        B. Filament ammeter
        C. Voltmeter
        D. Timer   Ref. 6 - p. 236

1162.   The major kilovoltage control knob varies the kilovoltage in steps of:
        A. 1 Kv
        B. 5 Kv
        C. 10 Kv
        D. 50 Kv   Ref. 6 - p. 241

1163.   To obtain a stereoscopic effect, a roentgenographic exposure must be
        obtained in what fashion?:
        A. With a high MaS
        B. Over a long arc
        C. From two different points of view
        D. On a rapid film changer   Ref. 6 - p. 365

1164.   The tube shift in stereoradiography is theoretically _____ the focus-
        film distance, when the viewing distance is 25 inches:
        A. 1/10
        B. 1/2
        C. Twice
        D. Four times   Ref. 6 - p. 366

1165.   The tube shift in stereoradiography is _____ to the dominant lines
        of the anatomic area:
        A. Parallel
        B. Oblique
        C. Adjacent
        D. At right angles   Ref. 6 - p. 366

1166.   In stereoradiography of the chest, the tube is shifted in what direction
        in relation to the ribs?:
        A. Crosswise
        B. At 45 degrees
        C. Parallel
        D. Crosswise or parallel   Ref. 6 - p. 366

1167.   If the focal film distance is to be 60 inches, the tube shift for a
        stereoradiograph will be:
        A. 30 inches
        B. 2 inches
        C. 0.6 inches
        D. 6 inches   Ref. 6 - p. 366

1168.   Between the two exposures of a stereo pair, the patient is:
        A. Moved
        B. Kept immobilized
        C. Recentered to the cassette
        D. Rotated 45 degrees   Ref. 6 - p. 367

1169.   In body section radiography, a structure having its long axis _____
        to the direction of motion is not completely blurred:
        A. Parallel
        B. Perpendicular
        C. Oblique   Ref. 6 - p. 382

1170.   When films are viewed stereoscopically, they are oriented on the
        viewer so as to represent a:
        A. Vertical tube shift
        B. Reversed image
        C. Horizontal tube shift
        D. Upside down image   Ref. 6 - p. 369

1171.   In body section radiography, the position of the fulcrum corresponds
        to the:
        A. Midpoint of the object
        B. Surface of the object
        C. Thickness of the object
        D. Objective plane   Ref. 6 - p. 386

1172.   In body section radiography, a separate film is exposed at the desired
        level:
        A. While the tube and Bucky tray are in motion
        B. While the tube is in motion but the Bucky tray is stationary
        C. While the tube is stationary but the Bucky tray is in motion
        D. While both the tube and Bucky tray are stationary
                                         Ref. 6 - p. 375

1173.   The simplest type of motion used in body section radiography is:
        A. Circular
        B. Polyhedral
        C. Spiral
        D. Rectilinear   Ref. 6 - p. 382

1174.   Multiple simultaneous body section roentgenograms are obtained
         by using A:
         A.   Polytome
         B.   Skip technique
         C.   "Book" cassette
         D.   Long tube travel   Ref. 6 - p. 387

1175.   Compared to film radiography, xeroradiography has lower:
         A.   Overall contrast
         B.   Detail visibility
         C.   Latitude
         D.   All of the above   Ref. 6 - p. 393

1176.   A photofluorographic unit is useful for:
         A.   Gastrointestinal radiology
         B.   Mass surveys of the chest
         C.   Urography
         D.   Myelography   Ref. 6 - p. 394

1177.   In photofluorography, the fluoroscopic image is recorded by a(an):
         A.   Motion picture camera
         B.   High speed camera
         C.   Video tape unit
         D.   Optical viewer   Ref. 6 - p. 394

1178.   The term definition, when applied to a roentgenogram, refers to the:
         A.   Sharpness of detail
         B.   Density differences
         C.   Degree of contrast
         D.   Degree of penetration   Ref. 6 - p. 298

1179.   A wide penumbra is manifested on a roentgenogram by:
         A.   Underpenetration of the center of the image
         B.   A blurred margin of the image
         C.   A white border at the film corners
         D.   A black border around the film   Ref. 6 - p. 300

1180.   Sharpness of the radiographic image is improved by all but one of the
         following:
         A.   A smaller effective focus
         B.   An increased focal-film distance
         C.   Decreased object-film distance
         D.   Increased object-film distance   Ref. 6 - p. 299

1181.   Motion unsharpness can be eliminated by:
         A.   Using long exposure times
         B.   Not using a compression band
         C.   Disregarding patient respiration
         D.   Careful immobilization of the part Ref. 6 - p. 302

1182.   In the 70 Kv range, the exposure can be doubled by an increase of:
         A.   6 Kv
         B.   10 Kv
         C.   20 Kv
         D.   50 Kv   Ref. 6 - p. 306

1183.   In the 30 Kv range, the exposure can be doubled by an increase of:
        A.   6 Kv
        B.   10 Kv
        C.   20 Kv
        D.   50 Kv   Ref. 6 - p. 306

1184.   When using a grid with a ratio of 8 or less, the kilovoltage should be
        kept below:
        A.   40 Kv
        B.   60 Kv
        C.   70 Kv
        D.   85 Kv   Ref. 6 - p. 306

1185.   For high voltage radiography (100-130 Kv), the grid ratio should be about:
        A.   4 to 1
        B.   6 to 1
        C.   8 to 1
        D.   12 to 1   Ref. 6 - p. 306

1186.   With an increase in the number of lead strips per inch in grid, the grid
        ratio is:
        A.   Decreased
        B.   Increased
        C.   Unchanged   Ref. 6 - p. 335

(T)RUE OR (F)ALSE:

1187.   Ionization is produced only by the addition or subtraction of orbital
        electrons.   Ref. 6 - p. 51

1188.   . In the process of induction the charged body induces the same kind of
        charge on the body placed in its field.   Ref. 6 - p. 57

1189.   Positive charges do not move readily in solid conductors.
                                            Ref. 6 - p. 59

1190.   A short, electrical conductor has more resistance to current flow than
        a long conductor.   Ref. 6 - p. 69

1191.   With increase in temperature of an electrical conductor resistance
        increases.   Ref. 6 - p. 69

1192.   Resistance within an electrical conductor is directly proportional to its
        cross sectional area.   Ref. 6 - p. 69

1193.   Current is equal in all parts of a parallel circuit.
                                            Ref. 6 - p. 78

1194.   Conductance in a parallel circuit is greater than in a series circuit.
                                            Ref. 6 - p. 77

1195.   When additional lights or other appliances are added to a parallel circuit
        the amperage in the main circuit decreases.
                                            Ref. 6 - p. 79

1196.   The voltage induced in the secondary coil of a transformer is dependent
        on the voltage applied to the primary coil and the ratio of the number of
        turns in the secondary coil to the number of turns in the primary coil.
                                            Ref. 6 - p. 121

1197.   In a step up transformer the current in the secondary coil is greater
         than that in the primary coil.   Ref. 6 - p. 122

1198.   An autotransformer operates on alternating current only.
                                                       Ref. 6 - p. 127

1199.   Control of voltage with a rheostat results in little heat loss.
                                                       Ref. 6 - p. 130

1200.   Rectifying systems in an X-ray circuit are located between the power
         supply and the primary coil of the transformer.
                                                       Ref. 6 - p. 139

1201.   The cloud of electrons formed around the incandescent filament of the
         cathode is known as a space charge.   Ref. 6 - p. 154

1202.   Brehms radiation constitutes 30 percent of emitted X-rays.
                                                       Ref. 6 - p. 157

1203.   Characteristic radiation consists of limited discrete wave lengths.
                                                       Ref. 6 - p. 157

1204.   The energy or penetrating power of an X-ray beam is directly propor-
         tioned to kilovoltage.   Ref. 6 - p. 157

1205.   Secondary radiation is emitted by atoms that have been struck by X-rays
                                                       Ref. 6 - p. 172

1206.   The characteristic radiation that results from photoelectric interaction
         is less energetic than the incident radiation and is absorbed in surround-
         ing tissue.   Ref. 6 - p. 177

1207.   Scattered radiation resulting from the Compton effect tends to deviate
         from the direction of the primary beam with increase of the kilovoltage
         of the primary beam.   Ref. 6 - p. 179

1208.   The phenomenon of pair production does not occur with kilovoltage levels
         ordinarily used in diagnostic radiology.   Ref. 6 - p. 180

1209.   Backscatter factor is influenced by source-skin distance.
                                                       Ref. 6 - p. 186

1210.   With increasing age of an X-ray tube, higher filament current is required
         for a desired milliamperage.   Ref. 6 - p. 200

1211.   Penumbra may be made smaller by decreasing focal-film distance.
                                                       Ref. 6 - p. 300

1212.   The most radiodense material in the body is bone.
                                                       Ref. 6 - p. 314

FOR EACH OF THE FOLLOWING MULTIPLE CHOICE QUESTIONS, CHOOSE THE ONE MOST APPROPRIATE ANSWER:

1213.   The ability of certain naturally occurring elements to undergo spontaneous nuclear disintegration and also elicit penetrating rays is known as:
A.   Natural radioactivity
B.   Artifical radioactivity
C.   Spontaneous disintegration
D.   Nuclear bombardment   Ref. 6 - p. 397

1214.   Artificially produced radioactive elements are known as:
A.   Inert elements
B.   Radiomimetic
C.   Radionuclides
D.   None of the above   Ref. 6 - p. 397

1215.   The naturally occurring radioactive elements can be grouped into three series: thorum, actinium and:
A.   Barium
B.   Uranium
C.   Iodine
D.   Iridium   Ref. 6 - p. 398

1216.   Alpha rays are:
A.   Helium ions
B.   Electrons
C.   Protons
D.   Photons   Ref. 6 - p. 398

1217.   Alpha particles have:
A.   1 negative charge
B.   2 negative charges
C.   1 positive charge
D.   2 positive charges   Ref. 6 - p. 398

1218.   Alpha rays are:
A.   Weakly ionizing
B.   Strongly ionizing
C.   Non-ionizing
D.   Moderately ionizing   Ref. 6 - p. 398

1219.   Beta rays are really:
A.   Photons
B.   Protons
C.   Neutrons
D.   High speed electrons   Ref. 6 - p. 399

1220.   Gamma rays are really:
A.   Photons
B.   Protons
C.   Light waves
D.   Electrons   Ref. 6 - p. 399

1221.   The ray with the largest mass is:
A.   Alpha
B.   Beta
C.   Gamma   Ref. 6 - p. 399

1222.   The penetrating ability of the beta rays is _____ than the alpha ray:
     A. Less
     B. Equal
     C. Greater   Ref. 6 - p. 399

1223.   Gamma rays are physically identical to:
     A. Radio waves
     B. X-rays
     C. Ultra violet rays
     D. Infra red rays   Ref. 6 - p. 399

1224.   Gamma rays originate in:
     A. Chemical bonds
     B. Magnetic fields
     C. The orbital rings
     D. Atomic nuclei   Ref. 6 - p. 399

1225.   The first breakdown product of radium is:
     A. Uranium
     B. Plutonium
     C. Radon
     D. Lead   Ref. 6 - p. 400

1226.   The stable end product of the breakdown of radium is:
     A. Radon
     B. Radium C
     C. Lead
     D. Uranium   Ref. 6 - p. 400

1227.   A sealed source of radium reaches a state of equilibrium with its breakdown products in about:
     A. 1 day
     B. 30 days
     C. 1 year
     D. 100 years   Ref. 6 - p. 400

1228.   A sealed source of radium emits:
     A. Alpha rays
     B. Beta rays
     C. Gamma rays
     D. All of the above   Ref. 6 - p. 400

1229.   The length of time required for 1/2 of the original amount of a radio-active substance to disintegrate is known as its:
     A. Transformation
     B. Count rate
     C. Half life
     D. Decay constant   Ref. 6 - p. 401

1230.   Radon is an element which under ordinary condition is a:
     A. Solid
     B. Liquid
     C. Gas
     D   Colloid   Ref. 6 - p. 402

1231.   If 20 milligrams of radium are applied to a tumor for 5 hours, the time intensity factor is:
     A. 100 milligram-hours
     B. 4 milligram-hours
     C. 100 milligram percent
     D. 4 millicuries   Ref. 6 - p. 403

1232. The unit for the time intensity factor when using radon therapy is the:
A. Milligram
B. Milligram hours
C. Gram
D. Millicurie   Ref. 6 - p. 403

1233. The radiation exposure at the site of tumor treated with radium depends on:
A. Time intensity factor
B. Distribution of the radium
C. Filtration of the applicator
D. All of the above   Ref. 6 - p. 404

1234. The applicators used in radium or radon treatment are made of metal; this metal acts as a(an):
A. Filter
B. Source of radiation
C. Insulator
D. Conductor   Ref. 6 - p. 405

1235. Any metal container for radium or radon will always absorb which type of radiation?:
A. Gamma
B. Beta
C. Alpha
D. Photons   Ref. 6 - p. 405

1236. The following types of containers are used for radium and radon:
A. Seeds
B. Needles
C. Tubes
D. All of the above   Ref. 6 - p. 406

1237. The best protection from radium rays is:
A. Lead gloves and apron
B. Gold or brass containers
C. Distance
D. Film badge   Ref. 6 - p. 406

1238. Isotopes of an element have the same chemical properties but have different:
A. Atomic numbers
B. Number of electrons
C. Number of protons
D. Mass numbers   Ref. 6 - p. 410

1239. When stable elements are converted to different radioactive elements by means of neutron capture, the process is known as:
A. Fusion
B. Transmutation
C. Fixation
D. Fission   Ref. 6 - p. 414

1240. The process(es) utilized in the production of radioisotopes is(are):
A. Transmutation
B. Fission
C. Radiative capture
D. All of the above   Ref. 6 - p. 415

1241.   Beta particles are detected indirectly due to their production of:
        A. Brehms radiation
        B. Heat
        C. Compton effect
        D. Photo electric effect   Ref. 6 - p. 416

1242.   If the half life of a radioactive element, having an activity of 200 mc.
        is 12 days, how much remains after 24 days?:
        A. 50 mc
        B. 5 mc
        C. 100 mc
        D. 75 mc   Ref. 6 - p. 418

1243.   If 10 grams of a stable element are bombarded in a nuclear reactor and
        the activity of the resultant radioactive isotope is 40 curies, then the
        specific activity of this sample is:
        A. 400 curies per gram
        B. 4 curies per gram
        C. 1/4 curies per gram
        D. 400 gram-curies   Ref. 6 - p. 416

1244.   The rate of decay of a radioactive isotope:
        A. Varies with its age
        B. Varies with the temperature
        C. Is constant
        D. Varies with the thickness of its container
                                           Ref. 6 - p. 416

1245.   The effective half life of a radioactive isotope is a function of the:
        A. Physical half life only
        B. Biologic half life only
        C. Biologic and physical half life
        D. Age of the isotope   Ref. 6 - p. 419

1246.   The phenomenon of _____ is utilized in treatment of thyroid
        disease with I-131:
        A. Colloidal dispersion
        B. Selective absorption
        C. Differential absorption
        D. None of the above   Ref. 6 - p. 420

1247.   The Geiger-Muller counter may be damaged if operated in the
        _____ region:
        A. Plateau
        B. Continuous discharge
        C. Threshold   Ref. 6 - p. 424

1248.   The sensitivity of a counter and minimum activity that it can detect
        depends on:
        A. Radioisotope being counted
        B. Arrangement of source and counter
        C. Type of counter
        D. All of the above   Ref. 6 - p. 434

1249.   The well counter is really a modified:
        A. Geiger-Muller counter
        B. Scintillation counter
        C. Pocket dosimeter
        D. Oscilloscope   Ref. 6 - p. 426

1250. Two main types of recorders for radiation counters are:
A. Geiger-Muller and Scintillation
B. Well counter and probe
C. Pocket dosimeter and monitor
D. Scaler and rate meter   Ref. 6 - p. 426

1251. Reliability or reproducibility of a given count depends on:
A. Speed of recorded counts
B. Total number of recorded counts
C. Time elapsed during recording
D. Type of counter used   Ref. 6 - p. 429

1252. The short interval of time during which a radiation counter is unable to detect radiation is called:
A Dead or recovery time
B. Response time
C. Interval time
D. Reaction time   Ref. 6 - p. 433

1253. The ability of a counter to detect radiation is known as its:
A. Count rate
B. Responsibility
C. Activity
D. Efficiency   Ref. 6 - p. 433

1254. A diagnostic test that depends on uptake of a radioisotope is used for what organ of the body?:
A. Adrenal
B. Thyroid
C. Gall bladder
D. Heart   Ref. 6 - p. 439

1255. A diagnostic test that depends on excretion of a radioisotope by the body is used to detect:
A. Thyroid cancer
B. Masses in the brain
C. Pernicious anemia
D. Tuberculosis   Ref. 6 - p. 445

1256. An example of a radioisotope dilution test is the determination of:
A. Blood volume
B. Heart size
C. Kidney failure
D. Hyper-thyroidism   Ref. 6 - p. 446

1257. The therapeutic effect of I-131 is primarily the result of its emission of:
A. Gamma rays
B. Alpha particles
C Beta particles
D. Positrons   Ref. 6 - p. 451

1258. The advantages of cobalt-60 teletherapy over 200 Kv X-ray therapy are:
A. Improved percentage depth dose
B. Skin sparing effect
C. Reduced hazard of bone damage
D. All of the above   Ref. 6 - p. 452

1259. Filtration in a cobalt-60 irradiator is used to:
A. Make the beam more homogenous
B. Remove Alpha particles
C Remove Beta particles
D. All of the above   Ref. 6 - p. 453

1260. Radiation emitted by a SR-90 applicator consists of:
A. Gamma rays
B. Alpha particles
C. Beta particles
D All of the above   Ref. 6 - p. 454

1261. The isotope P-32 concentrates in:
A. Bone marrow, spleen and liver
B. Heart, lungs and skin
C. Brain, spinal cord and nerves
D. Stomach, intestines and gall bladder
                              Ref. 6 - p. 440

1262. The energy of the gamma rays from cobalt-60 is comparable to
_____ volts of X-ray:
A. 0.9 million
B. 1 million
C. 3 million
D. 2 million   Ref. 6 - p. 451

FOR EACH OF THE FOLLOWING MULTIPLE CHOICE QUESTIONS,
CHOOSE THE ONE MOST APPROPRIATE ANSWER:

1263.   Cerebral ventriculography entails putting _____ into the cranial
        bones:
        A.   Electrodes
        B.   Isotopes
        C.   Gas
        D.   Burr holes   Ref. 7 - p. 398

1264.   For pneumoencephalography, the patient is kept in the _____
        position while the air is being introduced:
        A.   Standing
        B.   Sitting
        C.   Prone
        D.   Supine   Ref. 7 - p. 398

1265.   For a pneumoencephalogram, air is usually put in through a(an):
        A.   Lumbar puncture
        B.   Burr hole
        C.   Vein
        D.   Artery   Ref. 7 - p. 398

1266.   The initial upright pneumoencephalograms outline the:
        A.   Lateral ventricles
        B.   Fourth ventricle
        C.   Third ventricle
        D.   All of the above   Ref. 4 - p. 786

1267.   With the patient supine, air inserted into the cerebral ventricular
        system will outline the:
        A.   Anterior horns of the lateral ventricles
        B.   Posterior horns of the lateral ventricles
        C.   Fourth ventricle
        D.   Spinal canal   Ref. 7 - p. 405

1268.   With the patient prone, air inserted into the cerebral ventricles will
        outline the:
        A.   Anterior horns of the lateral ventricles
        B.   Posterior horns of the lateral ventricles
        C.   Fourth ventricle
        D.   Spinal canal   Ref. 7 - p. 409

1269.   With the head in the lateral position, air inserted into the cerebral
        ventricles will outline the:
        A.   Ventricle on the side away from the film
        B.   Ventricle on the side closest to the film
        C.   Fourth ventricle
        D.   Spinal canal   Ref. 7 - p. 411

1270.   In the initial views obtained in pneumoencephalography the orbitomeatal
        line is _____ from the horizontal:
        A.   Depressed 25 degrees
        B.   Elevated 25 degrees
        C.   Depressed 5 degrees
        D.   Depressed 10 degrees   Ref. 4 - p. 786

1271.   In encephalography, the making of small side to side motions of the
        patient's head while obtaining a roentgenogram is called:
        A. Myelography
        B. Autotomography
        C. Stereography
        D. Stereotaxis   Ref. 7 - p. 416

1272.   Another recently developed method that can be used to detect
        abnormalities within the skull and brain is:
        A. Sialography
        B. Arteriography
        C. Ultrasound
        D. Scanography   Ref. 7 - p. 420

1273.   The type of contrast material most commonly used in myelography is:
        A. Air
        B. Absorbable
        C. Carbon dioxide
        D. Opaque   Ref. 7 - p. 422

1274.   In the brow up and brow down positions used in pneumoencephalography
        the orbitomeatal makes angle of _____ with the film cassette:
        A. 90 degrees
        B. 180 degrees
        C. 25 degrees
        D. 10 degrees   Ref. 4 - p. 790

1275.   The injection of contrast material during myelography is made with
        the patient in the _____ position:
        A. Horizontal
        B. Standing
        C. Erect
        D. Sitting   Ref. 7 - p. 423

1276.   Perirenal or retroperitoneal gas insufflation is used to study the:
        A. Pancreas
        B. Aorta
        C. Adrenal gland
        D. Colon   Ref. 7 - p. 510

1277.   The injection of air into the peritoneal cavity to study the intra-
        abdominal organs is called:
        A. Pneumoencephalogram
        B. Pneumothorax
        C. Pneumatocele
        D. Artificial pneumoperitoneum   Ref. 7 - p. 514

1278.   Intravenous cholangiography is used primarily to study the:
        A. Kidneys
        B. Biliary ducts
        C. Liver
        D. Gall bladder   Ref. 7 - p. 518

1279.   The biliary ducts are usually visible roentgenographically about
        _____ minutes following intravenous injection of the opaque media:
        A. 5
        B. 30
        C. 60
        D. 120   Ref. 7 - p. 523

1280. Following surgical removal of the gall bladder, the biliary ducts are often studied by injecting contrast material into:
A. A T-tube
B. The liver
C. A nasogastric tube
D. A colostomy   Ref. 7 - p. 534

1281. A helpful adjunct procedure during X-ray study of the biliary duct is:
A. Retrograde pyelography
B. Angiography
C. Barium enema
D. Tomography   Ref. 7 - p. 535

1282. A study performed to evaluate high blood pressure is the:
A. Rapid sequence IVP
B. Retrograde cystogram
C. Pulmonary angiogram
D. Transhepatic cholangiogram   Ref. 7 - p. 553

1283. An obstruction in the ureter may produce:
A. Hydrothorax
B. Hydronephrosis
C. Hypernephroma
D. Ptosis   Ref. 7 - p. 556

1284. The swallowing of a carbonated beverage is sometimes used in:
A. Barium enema exams
B. Small bowel exams
C. IVP in children
D. IVP in adults   Ref. 7 - p. 560

1285. The injection of opaque material into the abdominal aorta can be used to visualize the:
A. Renal arteries
B. Carotid arteries
C. Subclavian arteries
D. Aortic valve   Ref. 7 - p. 560

1286. Retrograde pyelography does not demonstrate:
A. Kidneys
B. Ureters
C. Kidney function
D. Urinary stones   Ref. 7 - p. 562

1287. Retrograde pyelography is accomplished by:
A. Insertion of a catheter into the ureter
B. I-V injection of contrast material
C. Filling the bladder with contrast material
D. Inserting a catheter into the femoral artery
                                   Ref. 7 - p. 562

1288. Hysterosalpingography is performed to evaluate the:
A. Placental site
B. Patency of the uterine tubes
C Female urethra
D. Fetal age   Ref. 7 - p. 576

1289. A finding often looked for in hysterosalpingography is spill of opaque material into the:
   A. Peritoneal cavity
   B. Placenta
   C. Pleural cavity
   D. Duodenum   Ref. 7 - p. 577

1290. In a breech presentation, the fetal pelvis is directed toward the:
   A. Mother's head
   B. Mother's pelvis
   C. Mother's front
   D. Mother's back   Ref. 7 - p. 578

1291. Vertex presentation means that the fetal head is directed:
   A. Downward
   B. Upward
   C. To the side   Ref. 7 - p. 578

1292. The dimensions of the pelvic inlet and outlet can be determined by performing:
   A. Amniography
   B. Pelvimetry
   C. Pyelography
   D. Tomography   Ref. 7 - p. 596

1293. The diagnosis of tumors of the female breast is aided by the use of:
   A. Placentography
   B. Radiation therapy
   C. Mammography
   D. Bone survey   Ref. 7 - p. 602

1294. Three views commonly used in soft tissue radiography of the female breast are:
   A. Supero-inferior, medio-lateral, axillary
   B. Supero-inferior, right and left oblique
   C. Horizontal, erect and lateral
   D. Medio-lateral, right and left oblique
                                   Ref. 7 - p. 603

1295. The exposure technique in soft tissue radiography of the female breast entails the use of:
   A. Fine grain industrial film
   B. 25-35 Kv
   C. Short FFD
   D. All of the above   Ref. 7 - p. 604

1296. The opaque material generally used in angiography is a(an):
   A. Water soluble organic iodide
   B. Oily organic iodine
   C. Barium preparation
   D. Lead salt   Ref. 7 - p. 614

1297.- The opaque material in angiography may be introduced:
   A. Percutaneously
   B. By surgically exposing the blood vessel
   C. By selective catherization
   D. All of the above   Ref. 7 - p. 614

1298.   A device commonly used in angiography is the:
        A. Portable X-ray unit
        B. Rapid film changer
        C. Erect bucky
        D. Kymograph   Ref. 7 - p. 614

1299.   Film changers in angiography may be of the:
        A. Roll film type
        B. Cut film type
        C. Biplane type
        D. All of the above   Ref. 7 - p. 614

1300.   Cerebral angiography is performed by injecting opaque medium into the:
        A. Heart
        B. Carotid artery
        C. Abdominal aorta
        D. Femoral artery   Ref. 7 - p. 616

1301.   The amount of opaque material used for each injection in cerebral
        angiography is about:
        A. 10 cc
        B. 40 cc
        C. 75 cc
        D. More than 75 cc   Ref. 7 - p. 616

1302.   The injection method used in cerebral angiography is the:
        A. Intravenous
        B. Subcutaneous
        C. Percutaneous
        D. Intramuscular   Ref. 7 - p. 617

1303.   Positioning for cerebral angiography may include the following view:
        A. Lateral
        B. Antero-posterior
        C. Half axial and/or oblique
        D. All of the above   Ref. 7 - p. 618

1304.   Injection of opaque material into the vein at the elbow, or into the right
        atrium is used for:
        A. Cardiac angiography
        B. Carotid angiography
        C. Renal angiography
        D. Femoral angiography   Ref. 7 - p. 622

1305.   The translumbar method of injection is used to visualize the:
        A. Aortic arch
        B. Carotid arteries
        C. Abdominal aorta
        D. Thoracic aorta   Ref. 7 - p. 629

1306.   Insertion of a catheter into the femoral artery with passage into the
        abdominal aorta is an example of:
        A. Retrograde aortography
        B. Indirect aortography
        C. Intravenous aortography
        D. All of the above   Ref. 7 - p. 629

1307.   Local injection of opaque material into a renal artery is an example of:
        A.   Renal venography
        B.   Selective angiography
        C.   Intravenous aortography
        D.   All of the above   Ref. 7 - p. 633

1308.   Injection of opaque material directly into the spleen is called:
        A.   Splenic arteriography
        B.   Transhepatic cholangiography
        C.   T-tube cholangiography
        D.   Splenoportography   Ref. 7 - p. 633

1309.   Arteriography of the lower limb is obtained by injecting opaque contrast
        material into the:
        A.   Carotid artery
        B.   Brachial artery
        C.   Antecubital vein
        D.   Femoral artery   Ref. 7 - p. 635

1310.   In arteriography of the lower limb, the difference in density from the
        thigh to ankle can be compensated for by using:
        A.   Varying speed intensifying screens
        B.   A reciprocating bucky
        C.   High KV technique
        D.   Tomography   Ref. 7 - p. 637

1311.   For arteriography of the upper limb, the opaque medium can be injected
        into the:
        A.   Brachial artery
        B.   Carotid artery
        C.   Abdominal aorta
        D.   Femoral artery   Ref. 7 - p. 637

1312.   Lower extremity venography is useful for demonstrating:
        A.   Malignant tumors
        B.   Varicose veins
        C.   The heart and great vessels
        D.   Arterial obstruction   Ref. 7 - p. 638

1313.   In lower extremity venography, the table is usually:
        A.   Horizontal
        B.   Tilted head down
        B.   Tilted head up
        D.   Vertical   Ref. 7 - p. 638

1314.   A rapid film changer is not necessary for:
        A.   Pulmonary angiography
        B.   Renal angiography
        C.   Splenoportography
        D.   Lymphangiography   Ref. 7 - p. 640

1315.   Tomography is a method used for:
        A.   Taking radiographs of individual layers of the body
        B.   Evaluating leg length
        C.   The diagnosis of pregnancy
        D.   Routine chest radiography   Ref. 7 - p. 644

1316. In tomography, the height of the fulcrum is set to:
       A. The level of the layer desired
       B. Table top
       C. Front of the body
       D. Thickness of the part to be radiographed
                                          Ref. 7 - p. 644

1317. In tomography, the maximum exposure angle is obtained at the:
       A. Longest focal film distance
       B. Shortest focal film distance
       C. Intermediate focal film distance
       D. None of the above   Ref. 7 - p. 644

1318. In tomography of the lungs, the tube movement is set at:
       A. Right angles to the ribs
       B. Parallel to the ribs
       C. At 45 degrees to the ribs   Ref. 7 - p. 646

1319. Exposure for a tomogram is increased by about _____ percent as
       compared to a standard film of the same area:
       A. 5
       B. 20
       C. 50
       D. 75   Ref. 7 - p. 648

1320. In tomography, the objective plane is always placed:
       A. Perpendicular to the film
       B. Parallel to the film
       C. Obliquely to the film   Ref. 7 - p. 648

1321. Tomogram sections can be obtained at closer intervals if the:
       A. Exposure angle is increased
       B. Exposure angle is decreased
       C. Independent of the exposure angle   Ref. 7 - p. 649

1322. A small focal spot tube is necessary for:
       A. Serial angiography
       B. Magnification radiography
       C. Tomography
       D. Portable X-ray views   Ref. 7 - p. 664

1323. When the object-tube and object-film distances are equal, a(an)
       _____ image results:
       A. Enlarged
       B. Diminished
       C. Natural size   Ref. 7 - p. 664

1324. Stereographic films are obtained by exposing two separate films with:
       A. The object and film stationary
       B. The object and film stationary but changing the position of the tube
           focus
       C. Moving the position of the object
       D. Moving the position of the film   Ref. 7 - p. 682

1325. Stereographic films must be interpreted:
       A. Using monocular viewing
       B. Using binocular viewing
       C. Separately
       D. Superimposed on one another   Ref. 7 - p. 682

1326.   Stereographic exposures are made with the tube shift _____ to the
        long axis of the part to be examined:
        A. Parallel
        B. At right angles
        C. At 45 degrees
        D. Independent   Ref. 7 - p. 684

1327.   The base line used in cerebral angiography is the:
        A. Acanthomeatal line
        B. Orbitomeatal line
        C. Infraorbital meatal line
        D. Auricular line   Ref. 4 - p. 812

1328.   The gas ordinarily used in pelvic pneumography is:
        A. Oxygen
        B. Room air
        C. Nitrous oxide   Ref. 4 - p. 746

        FOR EACH OF THE FOLLOWING QUESTIONS, INSERT THE WORD
        WHICH MOST APPROPRIATELY MAKES THE ENTIRE STATEMENT
        ACCURATE AND COMPLETE:

1329.   In double contrast arthrography the contrast agents used are _____
        and _____.   Ref. 15 - p. 511

1330.   The temporal horns of the lateral ventricles are demonstrated in
        pneumoencephalography with the patient in the _____ position.
                                           Ref. 15 - p. 17

1331.   _____ tomography refers to the pneumoencephalographic technique
        in which the patient slowly rotates his head during exposure of a lateral
        film.   Ref. 15 - p. 27

1332.   Profile projections of nerve roots are obtained in lumbar myelography
        with the patient in the _____ position.
                                           Ref. 15 - p. 83

1333.   Injection of contrast material into the parotid gland is known as _____.
                                           Ref. 15 - p. 138

1334.   The contrast agent most often used in laryngography is _____.
                                           Ref. 15 - p. 144

1335.   Delivery of radiopaque medium to a specific heart chamber is called
        _____ angiography.   Ref. 15 - p. 218

1336.   Films should be obtained while respiration is suspended in the _____
        phase to better demonstrate esophageal varices.
                                           Ref. 15 - p. 359

1337.   In pneumoperitoneography the liver is outlined in the _____
        decubitus position.   Ref. - p. 515

        (T)RUE OR (F)ALSE:

1338.   There is little variability in the location of the gallbladder.
                                           Ref. 7 - p. 518

1339.   The normal gallbladder is not visible on a plain film of the abdomen.
                                           Ref. 7 - p. 519

1340. In lumbar myelography the patient is examined first in the prone position. Ref. 15 - p. 82

1341. It is always necessary to remove all contrast material after myelography. Ref. 15 - p. 84

1342. In the severely jaundiced patient oral cholecystography may not be successful. Ref. 7 - p. 522

1343. The oblique position in chest tomography is of particular value for mediastinal lesions. Ref. 7 - p. 654

1344. Opaque medium will not pass from the spinal subarachnoid space into the cerebral subarachnoid space. Ref. 7 - p. 424

1345. Discography refers to direct injection of opaque medium into an intervertebral disc. Ref. 7 - p. 427

1346. In most cases oblique projections are not required in bronchography. Ref. 7 - p. 471

1347. Fluoroscopy is usually performed in examination of the pharynx and esophagus. Ref. 7 - p. 481

1348. In the prone position the colon converges toward the midline. Ref. 7 - p. 496

1349. The gallbladder is located anterior to the coronal plane of the abdomen in the prone position. Ref. 7 - p. 535

1350. When compression is used in intravenous pyelography done for the study of hypertension, it is applied immediately before injection. Ref. 7 - p. 553

1351. The most suitable projections in mammography are the supero-inferior and the medio-lateral. Ref. 7 - p. 603

MATCH THE FOLLOWING GROUP OF SPECIAL PROCEDURES WITH THE APPROPRIATE ANATOMIC REGION LISTED IN THE RIGHT HAND COLUMN:

1352. ___ Femoral arteriogram	A.	Brain
1353. ___ Bronchogram	B.	Lungs
1354. ___ Carotid angiogram	C.	Spine
1355. ___ Angiocardiogram	D.	Uterus and fallopian tubes
1356. ___ Hysterosalpingogram	E.	Gall bladder
1357. ___ Arthrogram	F.	Arteries of the leg
1358. ___ Nephrotomogram	G.	Kidney
1359. ___ Cholangiogram	H.	Bile ducts
1360. ___ Myelogram	I.	Joints
1361. ___ Sialogram	J.	Salivary glands
	K.	Heart
	L.	Esophagus
	M.	Urethra
	N.	Femur
	O.	Colon

SELECT THE ONE MOST APPROPRIATE ANSWER:

1362.   A patient scheduled for an upper gastro-intestinal series should not have anything by mouth for at least _____ hours prior to the study:
   A. 2
   B. 6
   C. 24
   D. 48   Ref. 7 - p. 484

1363.   With the patient supine and the head extended, air in the cerebral ventricles will outline the:
   A. Posterior portion of the 3rd ventricle
   B. Superior portion of the 4th ventricle
   C. Posterior portion of the lateral ventricle
   D. Anterior inferior portion of the 3rd ventricle
                                       Ref. 4 - p. 792

1364.   If a film of the abdomen is taken approximately 24 hours after an upper GI series has been performed, the following structure may be demonstrated:
   A. Stomach
   B. Jejunum
   C. Duodenum
   D. Colon   Ref. 7 - p. 485

1365.   For the fluoroscopic examination of the stomach mucosa and duodenal bulb, the radiologist often uses the technique of:
   A. Delayed filming
   B. Compression
   C. Pneumoperitoneum
   D. Barium enema   Ref. 7 - p. 484

1366.   During the barium meal examination of the small bowel, films of the abdomen are obtained every 1/2-1 hour until:
   A. 3 hours have elapsed
   B. The patient complains of hunger
   C. The ileocecal valve is reached
   D. The sigmoid colon is reached   Ref. 7 - p. 491

1367.   Laxatives and enemas are used to prepare the usual patient for which examination?:
   A. Barium swallow
   B. Barium enema
   C. Chest fluoroscopy
   D. T-tube cholangiogram   Ref. 7 - p. 494

1368.   After completion of a barium enema, the patient is allowed to go to the toilet and is then returned for a(an):
   A. Post evacuation film
   B. Repeat enema
   C. Upper GI examination
   D. I. V. P.   Ref. 7 - p. 498

1369.   An additional method used to evaluate colon abnormalities is the:
   A. Hypaque enema
   B. Air-contrast technique
   C. "Coke" study
   D. Small bowel study   Ref. 7 - p. 500

1370.   Perirenal or retroperitoneal air studies are used to visualize the:
         A.   Stomach
         B.   Adrenal glands
         C.   Aorta
         D.   Liver   Ref. 7 - p. 510

1371.   The position of the gall bladder may be indicated on a plain film of the
         abdomen by the presence of:
         A.   Non-opaque calculi
         B.   Opaque calculi
         C.   Bile
         D.   Barium   Ref. 7 - p. 519

1372.   The opaque medium used for radiologic visualization of the gall bladder
         contains:
         A.   Organic iodine
         B.   Barium
         C.   Bismuth
         D.   Fat   Ref. 7 - p. 522

1373.   The effect of a fatty meal given at the end of an oral cholecystogram is to:
         A.   Dilate the gall bladder
         B.   Cause retention within the gall bladder
         C.   Cause contraction of the gall bladder
         D.   Stimulate the pancreas   Ref. 7 - p. 522

1374.   On an intravenous cholangiogram, the bile ducts are normally best
         visualized within:
         A.   3 hours
         B.   60-75 minutes
         C.   1-1/2 - 2 hours
         D.   25-30 minutes   Ref. 7 - p. 522

1375.   On an intravenous cholangiogram, the gall bladder is best visualized
         within:
         A.   25-30 minutes
         B.   2 hours
         C.   8 hours
         D.   24 hours   Ref. 7 - p. 522

1376.   Before an intravenous urogram is performed, the patient is instructed
         not to drink fluids for:
         A.   2 hours
         B.   12 hours
         C.   24 hours
         D.   48 hours   Ref. 7 - p. 540

1377.   Compression is used during an intravenous urogram to:
         A.   Help excretion
         B.   Immobilize the patient
         C.   Prevent patient breathing
         D.   Retain the contrast material in the kidneys
                                                 Ref. 7 - p. 540

1378.   Films taken during an intravenous urogram must be labeled as to:
         A.   Time obtained
         B.   Time of day
         C.   Phase of respiration
         D.   Top and bottom   Ref. 7 - p. 542

1379.   For a retrograde pyelogram, the opaque material is inserted:
A.   Through a ureteral catheter
B.   Into a vein
C.   By mouth
D.   By enema   Ref. 7 - p. 542

1380.   Rapid film changers are used primarily in:
A.   Cholecystography
B.   Urography
C.   Angiography
D.   Pelvimetry   Ref. 7 - p. 615

1381.   When performing a lymphangiogram, second set of films are obtained in:
A.   2 hours
B.   24 hours
C.   1 week
D.   1 month   Ref. 7 - p. 640

1382.   A tomogram differs from a standard radiograph in that:
A.   Less contrast is obtained
B.   Only one plane is in focus
C.   All motion is recorded
D.   Different type film is used   Ref. 7 - p. 644

1383.   In tomography, the X-ray tube moves in a plane _____ to the film:
A.   Perpendicular
B.   At right angles
C.   Parallel
D.   Adjacent   Ref. 7 - p. 644

1384.   The exposure necessary to obtain a tomogram is increased by about
_____ percent as compared with a standard film of the same part:
A.   5
B.   20
C.   50
D.   100   Ref. 7 - p. 648

1385.   Tomograms can be taken at closer section intervals if the:
A.   Kv is increased
B.   Kv is decreased
C.   Exposure angle is decreased
D.   Exposure angle is increased   Ref. 7 - p. 649

1386.   The majority of lung tomograms are taken with the patient in the
_____ position:
A.   Apical lordotic
B.   Decubitus
C.   Erect
D.   Supine   Ref. 7 - p. 651

1387. To produce a radiograph which possesses depth, one uses the technique of:
A. Stereography
B. Magnification
C. Tomography
D. Cineradiography   Ref. 7 - p. 682

1388. In order to produce such a "depth" roentgenogram, two separate films are exposed and:
A. The position of the tube focus is changed
B. The focal film distance is changed
C. Contrast media is injected
D. The patient is moved between exposures
Ref. 7 - p. 682

FOR EACH OF THE FOLLOWING MULTIPLE CHOICE QUESTIONS, CHOOSE THE ONE MOST APPROPRIATE ANSWER:

1389.   The portion of the eye which receives impulses of light and transfers this impression to the brain is the:
A. Lens
B. Pupil
C. Retina
D. Sclera   Ref. 8 - p. 1

1390.   Effective dark adaptation depends on the sensitivity of the _____ in dim light:
A. Optic nerves
B. Rods
C. Vestibular apparatus
D. Sclera   Ref. 8 - p. 2

1391.   During fluoroscopy, the X-ray intensity at the fluoroscopic screen is only a fraction of the intensity emitted at the tube because of:
A. Absorption by the patient's body
B. Scatter
C. Diffusion
D. Absorption by the fluoroscopist's body
           Ref. 8 - p. 7

1392.   Photo cathodes emit _____ when exposed to light:
A. Protons
B. X-rays
C. Electrons
D. Neutrons   Ref. 8 - p. 10

1393.   Photo cathodes are usually made of _____ metals:
A. Alkaline
B. Ferrous
C. Acidic   Ref. 8 - p. 10

1394.   The gain in brightness due to electronic intensification is:
A. 40-50
B. 12-15
C. 5-10
D. 100-150   Ref. 8 - p. 13

1395.   Resolution of an intensifier tube is:
A. Greater in the central region of the tube
B. Greater in the periphery of the tube
C. Uniform over the face of the tube   Ref. 8 - p. 14

1396.   Decrease in the quantity of the X-ray reaching the input phosphor results in _____ in scintillation:
A. Decrease
B. Increase
C. No change   Ref. 8 - p. 18

1397.   Loss of brightness occurs in all image intensifier tube output phosphors. This light "fall off" occurs at which portion of the tube?:
A. Center
B. Cathode
C. Anode
D. Periphery   Ref. 8 - p. 20

1398.   Image intensifier tube size (5 inch, 7 inch, etc.) refers to the:
        A. Size of the face of the tube
        B. Length of the tube
        C. Field size covered
        D. Focal spot size   Ref. 8 - p. 21

1399.   When doing image intensification work, as small a focal spot size as
        possible should be used. The factor limiting the size of the focal spot
        that may safely be used is the:
        A. Rating of the tube
        B. Magnification
        C. Line voltage
        D. Glass envelope   Ref. 8 - p. 22

1400.   During image intensified fluoroscopic procedures lights over the table
        and in back of the operator must be:
        A. Dimmed
        B. Left on
        C. Turned off   Ref. 8 - p. 36

1401.   Cinefluorography is the taking of:
        A. Fluoroscopic spot films
        B. Motion pictures of the fluoroscopic image
        C. Photo-timed exposures
        D. Photofluorograms   Ref. 8 - p. 43

1402.   In cinefluorography, a mirror is used to deflect a small percent of the
        light to the viewing system, and the remainder of the light to the:
        A. Intensifier tube
        B. Eye
        C. Camera
        D. Radiologist   Ref. 8 - p. 43

1403.   The number of individual pictures per unit time taken by a motion picture
        camera is called:
        A. Frames per second
        B. Feet per second
        C. Ergs per second
        D. Lines per millimeter   Ref. 8 - p. 47

1404.   Cine film is usually the _____ type:
        A. Panchromatic
        B. Orthochromatic
        C. Non-color sensitized   Ref. 8 - p. 60

1405.   Desirable features(s) of cine projector is(are):
        A. Single framing
        B. Instantaneous forward and reverse motions
        C. Flicker free projection
        D. All of the above   Ref. 8 - p. 68

1406.   To obtain natural motion of the subject being studied, the cine projector
        should be run at _____ frame rate as that used in the cine camera:
        A. A faster
        B. The same
        C. A slower   Ref. 8 - p. 68

1407.   In synchronized cineradiographic systems frame rates are multiples of:
        A.   The electrical source frequency
        B.   RPM of the camera drive
        C.   Line voltage
        D.   All of the above   Ref. 8 - p. 71

1408.   Frame rate in synchronized cineradiographic systems may be _____
        frames per second:
        A.   7-1/2
        B.   15
        C.   30
        D.   All of the above   Ref. 8 - p. 71

INDICATE IN EACH OF THE FOLLOWING STATEMENTS WHETHER
(T)RUE OR (F)ALSE:

1409.   With the eyes darked adapted and illumination levels used in fluoroscopy
        contrast discrimination is good.   Ref. 8 - p. 4

1410.   Materials forming the photocathode of an image intensifier emit electrons
        in direct proportion to the amount of light striking them.
                                                  Ref. 8 - p. 4

1411.   The image on the output phosphor is reduced in size and inverted.
                                                  Ref. 8 - p. 13

1412.   The size of an image formed by a lens is directly proportional to its
        focal length   Ref. 8 - p. 37

1413.   The vidicon television camera tube is larger and more complex than the
        image orthicon tube.   Ref. 8 - p. 55

1414.   Panchromatic film is sensitive only to red and infrared light.
                                                  Ref. 8 - p. 58

1415.   Cine film is a single emulsion film.   Ref. 8 - p. 67

1416.   Synchronized cine systems with grid controlled X-ray tube reduce
        patient dosage to one half that of non-synchronized systems.
                                                  Ref. 8 - p. 76

1417.   The increase in patient dosage when filming with 35 mm. camera is a
        function of the size of the film.   Ref. 8 - p. 83

1418.   Automatic brightness control systems function to maintain the brightness
        of an image by changes in milliamperage of the X-ray tube.
                                                  Ref. 8 - p. 84

1419.   The 5 inch intensifier tube is ideal for examinations such as upper
        gastrointestinal series.   Ref. 8 - p. 22

1420.   The X-ray exposure dose to the patient is increased when one uses image
        intensification instead of conventional fluoroscopy.
                                                  Ref. 7 - p. 697

1421.   The current that is necessary for image intensification viewing ranges
        between 0.2-2.0 milliamperes.   Ref. 7 - p. 697

1422.   Because of the high milliamperage required with image intensification, a fine focus tube can not be used.   Ref. 7 - p. 697

1423.   Biplane fogging in cinefluorography can be eliminated by using grids and properly localizing the X-ray beam in both planes.
Ref. 7 - p. 698

1424.   Cine speeds of 8 to 16 frames per second are too slow for the usual upper gastrointestinal examination.   Ref. 7 - p. 698

1425.   Cine films are available in both 16 mm. and 35 mm. widths.
Ref. 7 - p. 699

COMPLETE THE FOLLOWING STATEMENTS:

1426.   Red goggles are used by the fluoroscopist for _____.
Ref. 8 - p. 2

1427.   The rods of the retina contain a pigment known as _____.
Ref. 8 - p. 2

1428.   Color perception depends upon the _____ cells in the retina.
Ref. 8 - p. 2

1429.   Objects that block off light completely are called _____.
Ref. 8 - p. 5

1430.   Light intensity at any point is known as _____.
Ref. 8 - p. 6

1431.   The law which states that: "Illumination varies inversely as the square of the distance of that point from the light source," is called the _____ law.   Ref. 8 - p. 6

1432.   The basic component of the image intensifier is an _____.
Ref. 8 - p. 10

1433.   "Lines per inch" is an expression of _____.
Ref. 8 - p. 13

1434.   Another name for scintillation is _____.
Ref. 8 - p. 17

1435.   The loss of light at the edges of the image as compared with the center of the image is known as _____.   Ref. 8 - p. 20

1436.   The under table tube usually used in image intensification has a fixed table top to target distance of _____.
Ref. 8 - p. 21

1437.   The largest focal spot that should be used in an image intensification unit is _____.   Ref. 8 - p. 22

1438.   An objective and field lens are found in the _____.
Ref. 8 - p. 31

1439.   The f number of a lens measures its _____.
Ref. 8 - p. 38

1440.   The smaller the f number of the lens, the _____ light is transmitted through that lens.   Ref. 8 - p. 38

1441.   Adjusting the distance between the lens and the film is known as _____.
        Ref. 8 - p. 39

1442.   When combining cinefluorography with image intensification, a mirror is used to deflect about _____ percent of the light to the camera.
        Ref. 8 - p. 43

1443.   The number of individual pictures taken per second during a cine examination is referred to as _____.   Ref. 8 - p. 47

1444.   A 16 mm. camera takes _____ frames per foot of film.
        Ref. 8 - p. 48

1445.   A 35 mm. camera takes _____ frames per foot of film.
        Ref. 8 - p. 49

1446.   High speed, high contrast film can be used in cinefluorography, but _____ is then sacrifieced.   Ref. 8 - p. 57

1447.   The ASA number of a type of cine film indicates its _____.
        Ref. 8 - p. 63

1448.   Negative films are also known as _____.
        Ref. 8 - p. 64

1449.   Cine film should be developed for _____ speed and contrast.
        Ref. 8 - p. 67

FOR EACH OF THE FOLLOWING MULTIPLE CHOICE QUESTIONS,
CHOOSE THE ONE MOST APPROPRIATE ANSWER:

1450.   Natural background radiation refers to:
    A.   Cosmic rays
    B.   Naturally radioactive minerals
    C.   Radioactive isotopes normally in the body tissues
    D.   All of the above   Ref. 6 - p. 458

1451.   The accumulated occupational maximum permissible dose from X,
    Gamma or Beta radiation to the whole body is calculated by using the
    formula:
    A.   MPD = 5 (n -18) rads
    B.   MPD = 5N -18 rads
    C.   MPD = 5 (18 + N) rads   Ref. 6 - p. 460

1452.   The annual maximal permissible dose for the general population is
    _____ that for radiation workers:
    A.   1/10
    B.   1/4
    C.   1/2
    D.   Twice   Ref. 6 - p. 461

1453.   When exposure levels are below the weekly maximal permissible dose,
    the film badge worn by a radiation worker may be replaced every:
    A.   Week
    B.   Month
    C.   Six months
    D.   Year   Ref. 6 - p. 464

1454.   Adequate dark adaptation for fluoroscopic work is obtained in:
    A.   1 minute
    B.   5 minutes
    C.   30 minutes
    D.   3 hours   Ref. 6 - p. 474

1455.   Proper factors for fluoroscopy would be:
    A.   145 Kv and 4 Ma
    B.   90 Kv and 50 Ma
    C.   85 Kv and 4 Ma
    D.   130 Kv and 40 Ma   Ref. 6 - p. 475

1456.   The best protection from a radium source is:
    A.   Aluminum and copper
    B.   Lead and distance
    C.   Glass
    D.   Lead gloves and apron   Ref. 6 - p. 480

1457.   Radium should always be picked up:
    A.   With the fingers
    B.   With rubber gloves
    C.   With lead gloves
    D.   With an instrument such as forceps   Ref. 6 - p. 481

1458.   The most satisfactory shielding material is usually:
    A.   Barium
    B.   Lead
    C.   Plaster
    D.   Wood   Ref. 6 - p. 485

1459.   An example of a radiation survey instrument is the:
         A.   "Cutie pie"
         B.   Film badge
         C.   Pocket dosimeter
         D.   Well counter   Ref. 6 - p. 487

1460.   The somatic effects of radiation refer to those experienced by which
         of the following groups?:
         A.   Unborn generations
         B.   The exposed individual
         C.   The un-exposed individual
         D.   None of the above   Ref. 7 - p. 752

1461.   Some parts of the body are extremely sensitive to radiation, such as the:
         A.   Gonads
         B.   Blood forming organs
         C.   Lens of the eye
         D.   All of the above   Ref. 7 - p. 752

1462.   High doses of radiation to pregnant women can result in:
         A.   Miscarriage
         B.   Stillbirth
         C.   Developmental abnormalities in the child
         D.   All of the above   Ref. 7 - p. 752

1463.   One method that is useful to keep radiation exposure to a minimum is:
         A.   The inverse square law
         B.   Using the largest possible X-ray beam
         C.   Standing as close as possible to the table during fluoroscopy
         D.   Never wearing a lead protective apron
                                                    Ref. 7 - p. 754

1464.   Gonad shields of 0.5 mm of lead will reduce the gonad dose to about:
         A.   5%
         B.   25%
         C.   50%
         D.   75%   Ref. 7 - p. 755

1465.   For a given film dose, the dose to the skin is _____ for a long
         focal-film distance than for a short focal-film distance:
         A.   The same
         B.   More
         C.   Less   Ref. 7 - p. 755

1466.   Artificial background radiation arises from:
         A.   Cosmic rays
         B.   Medical X-rays
         C.   Radioactive minerals
         D.   All of the above   Ref. 6 - p. 459

1467.   Radiation induced cataracts are characterized as a _____ effect
         of radiation exposure:
         A.   General
         B.   Local
         C.   Genetic
         D.   Regional   Ref. 6 - p. 465

1468.   The dose rate at the table top in fluoroscopy should be less than:
        A. 10 R per hour
        B. 10 R per minute
        C. 1 R per minute
        D. 1 R per hour   Ref. 6 - p. 475

1469.   Radiation coming directly from the target of an X-ray tube is called:
        A. Secondary
        B. Scattered
        C. Stray
        D. Primary   Ref. 7 - p. 758

1470.   That portion of the primary beam which passes through the collimating
        device is called:
        A. Useful beam
        B. Scattered beam
        C. Absorbed beam
        D. None of the above   Ref. 7 - p. 758

1471.   The maximum permissible dose that may be accumulated in one year due
        to occupational exposure is:
        A. 1 rads
        B. 25 rads
        C. 5 rads
        D. 50 rads   Ref. 6 - p. 461

1472.   The maximum occupational dose that may be accumulated by radiation
        workers in any 13 week period is:
        A. 13 rads
        B. 75 rads
        C. 3 rads
        D. 25 rads   Ref. 6 - p. 461

1473.   Radiation emitted by matter irradiated with X-rays is called:
        A. Stray radiation
        B. Secondary radiation
        C. Absorbed radiation
        D. All of the above   Ref. 7 - p. 758

1474.   The most precise personal monitoring device is:
        A. Film badge
        B. Pocket dosimeter
        C. Blood counts   Ref. 6 - p. 463

1475.   Under usual conditions the most convenient and accurate method of
        personal monitoring is:
        A. Ring dosimeter
        B. Pocket dosimeter
        C. Film badge   Ref. 6 - p. 463

1476.   The minimum level of radiation exposure below which no genetic or
        somatic damage occurs is:
        A. Not established
        B. 25 rads
        C. 50 rads   Ref. 4 - p. 419

1477.   Lead aprons worn during fluoroscopy should be equivalent to at least
         _____ mm of lead:
         A. 0.05
         B. 0.5
         C. 5   Ref. 4 - p. 421

1478.   A woman undergoing examination of the lumbar spine may receive a
         gonadal dose of as much as _____ rads per film taken:
         A. 0.27
         B. 27
         C. 270   Ref. 4 - p. 424

1479.   In fluoroscopic equipment the distance between the fluoroscopic tube
         and the table top should be not less than:
         A. 10 inches
         B. 8 inches
         C. 15 inches   Ref. 11 - p. 7

1480.   Fluoroscopic exposure switches must be _____ type:
         A. Toggle
         B. Dead man
         C. Button   Ref. 11 - p. 8

1481.   The cumulative timing device on fluoroscopic equipment must indicate a
         period of irradiation by:
         A. Audible signal
         B. Interruptions of exposure
         C. Either of the above   Ref. 11 - p. 9

1482.   Mobile fluoroscopic equipment must be provided with:
         A. Image intensification
         B. Television monitoring
         C. Explosion proof circuits   Ref. 11 - p. 9

1483.   Milliamperage and kilovoltage employed in cineradiography are _____
         those used in fluoroscopy:
         A. Less than
         B. The same as
         C. Higher than   Ref. 11 - p. 12

1484.   The exposure switch on mobile radiographic equipment must be
         arranged so that the operator can stand at least _____ feet away
         from the useful beam:
         A. 12
         B. 6
         C. 4   Ref. 11 - p. 16

1485.   Film badges should _____ be worn by a radiation worker when
         undergoing medical or dental X-ray examinations:
         A. Always
         B. Never
         C. Occasionally   Ref. 11 - p. 33

1486.   If a lead apron is worn, the monitoring device should be worn:
         A. Outside of the apron
         B. Under the apron
         C. On the wrist
         D. Need not be worn at all   Ref. 9 - p. 10

1487. Protective gloves should be of at least _____ lead equivalent:
A. 1 mm
B. 0.25 mm
C. 5 mm
D. 10 mm   Ref. 9 - p. 13

1488. When examining a patient in the reproductive age group, special care should be taken to avoid exposure to the:
A. Eye
B. Extremities
C. Chest
D. Gonads   Ref. 9 - p. 13

1489. Primary radiation barriers in walls should be how high?:
A. 2 feet
B. 5 feet
C. 7 feet
D. None of the above   Ref. 9 - p. 14

1490. Transmission of X-rays through thick protective barriers is closely related to:
A. Peak operating potential of the X-ray tube
B. Milliamperage
C. Exposure time
D. Added filtration   Ref. 9 - p. 42

1491. Primary barriers protect against:
A. Secondary radiation
B. Scattered radiation
C. Leakage radiation
D. Radiation of the useful beam   Ref. 9 - p. 42

1492. Secondary barriers are those which are exposed only to:
A. Leakage radiation
B. Scattered radiation
C. Both of the above
D. None of the above   Ref. 9 - p. 42

1493. The target to skin distance in mobile diagnostic X-ray equipment should be not less than:
A. 1 foot
B. 3 feet
C. 6 feet
D. 6 inches   Ref. 9 - p. 15

1494. Where the workload is sufficiently low in a dental radiographic unit, shielding is not necessary if the:
A. Unit is well collimated
B. Operator stands at least 6 feet away
C. Kv is less than 70
D. Filtration is adequate   Ref. 9 - p. 17

1495. "Grenz ray" refers to X-rays produced at potentials below:
A. 150 Kv
B. 300 Kv
C. 50 Kv
D. 20 Kv   Ref. 9 - p. 20

1496.   When using Grenz rays, the operator need not be shielded unless he is
         exposed to a target-skin distance of less than:
         A. 6 inches
         B. 1 foot
         C. 3 meters
         D. 6 meters   Ref. 9 - p. 20

1497.   "Contact" therapy uses a potential of about:
         A. 10-20 Kv
         B. 40-50 Kv
         C. 60-80 Kv
         D. 90 Kv   Ref. 9 - p. 20

1498.   Leakage radiation at the surface of a therapy tube which is operated
         below 60 Kv should not be greater than what value?:
         A. 1 r/hour
         B. 0.1 r/hour
         C. 10 r/hour
         D. 1 mr/hour   Ref. 9 - p. 21

1499.   Movable protective screens should not be used with equipment operating
         above what potential?:
         A. 50 Kv
         B. 70 Kv
         C. 90 Kv
         D. 125 Kv   Ref. 9 - p. 19

1500.   X-ray therapy apparatus constructed with beryllium or other low-
         filtration windows may produce a dose rate of more than _____
         r/min at the aperture:
         A. 10
         B. 100
         C. 1000
         D. 1,000,000   Ref. 9 - p. 21

1. Young, C. G. and Likos, J. J.: "Medical Specialty Terminology." Volume 2, C. V. Mosby Company, St. Louis, Missouri, 1972.

2. Dorland's Illustrated Medical Dictionary, 24th edition, W. B. Saunders Company, Philadelphia, Pennsylvania, 1965.

3. Mallet, M.: "A Handbook of Anatomy and Physiology for Student X-ray Technicians," 4th Edition, American Society of Radiologic Technologists, Fond du Lac, Wisconsin, 1966.

4. Merril, V.: "Atlas of Roentgenographic Positions," 3rd Edition, C. V. Mosby Company, St. Louis, Missouri, 1967.

5. Fuchs, A. W.: "Principles of Radiographic Exposure and Processing," 2nd Edition, Charles C Thomas, Springfield, Illinois, 1958.

6. Selman, J.: "Fundamentals of X-ray and Radium Physics," 5th Edition, Charles C Thomas, Springfield, Illinois, 1972.

7. Clark, K. C.: "Positioning in Radiography," 8th Edition, Grune and Stratton, New York, New York, 1964.

8. "Image Intensification and Recording Principles," X-ray Department, General Electric Corporation, Milwaukee, Wisconsin, 1963.

9. Handbook 76: "Medical X-ray Protection up to Three Million Volts," U. S. Department of Commerce, N. B. S., Washington, D. C., 1961.

10. "The Fundamentals of Radiography," 9th Edition, Medical Division, Eastman Kodak Company, Rochester, New York.

11. NCRP Report No. 33: "Medical X-ray and Gamma-Ray Protection for Energies up to 10 MeV," National Council on Radiation Protection and Measurements, Washington, D. C., 1968.

12. "X-Omat" Handbook, Eastman Kodak Company, Rochester, New York.

13. van der Plaats, G. L.: "Medical X-ray Technique," 3rd Edition, Macmillan Company, New York, New York, 1972.

14. Cullinan, J. E.: "Illustrated Guide to X-ray Technics." J. B. Lippincott Company, Philadelphia, Pennsylvania, 1972.

15. Beranbaum, S. L. and Meyers, P. H.: "Special Procedures in Roentgen Diagnosis," Charles C Thomas, Springfield, Illinois, 1964.

ANSWER KEY

The Authors have made every effort to thoroughly verify the questions and answers. In a volume of this size some inaccuracies and ambiguities may occur. Therefore, if in doubt, consult your references.

THE PUBLISHERS

SECTION I

1. B	47. A	93. C	139. D	182. D	228. D
2. A	48. C	94. B	140. A	183. B	229. C
3. A	49. B	95. A	141. B	184. D	230. D
4. C	50. A	96. B	142. C	185. B	231. B
5. D	51. C	97. C	143. D	186. D	232. D
6. B	52. D	98. D	144. A	187. C	233. D
7. D	53. B	99. C	145. B	188. B	234. C
8. C	54. B	100. A	146. C	189. A	235. A
9. A	55. B	101. A	147. B	190. D	236. C
10. B	56. C	102. D	148. D	191. A	237. A
11. A	57. B	103. C	149. A	192. B	238. B
12. C	58. D	104. A	150. C	193. C	239. C
13. D	59. B	105. D	151. B	194. B	240. D
14. A	60. A	106. B	152. D	195. D	241. B
15. D	61. A	107. C	153. A	196. D	242. C
16. B	62. D	108. A	154. B	197. A	243. D
17. A	63. D	109. C	155. D	198. B	244. D
18. B	64. C	110. B	156. C	199. A	245. B
19. A	65. A	111. D	157. A	200. B	246. A
20. D	66. C	112. C	158. A	201. C	247. D
21. C	67. C	113. B	159. B	202. D	248. C
22. B	68. A	114. C	160. D	203. C	249. B
23. A	69. B	115. C	161. C	204. D	250. D
24. B	70. B	116. B	162. A	205. B	251. C
25. D	71. A	117. A	163. D	206. A	252. B
26. C	72. D	118. D	164. B	207. C	253. A
27. C	73. A	119. C	165. D	208. A	254. A
28. A	74. D	120. C	166. A	209. D	255. B
29. D	75. B	121. B	167. C	210. B	256. B
30. A	76. D	122. A	168. C	211. C	257. B
31. C	77. C	123. D	169. D	212. D	258. C
32. D	78. B	124. C	170. A	213. B	259. A
33. B	79. C	125. B	171. B	214. A	260. D
34. A	80. D	126. D	172. A	215. D	261. B
35. B	81. B	127. D	173. C	216. B	262. B
36. A	82. B	128. C	174. B	217. D	263. C
37. B	83. A	129. B	175. A	218. C	264. B
38. D	84. C	130. C		219. C	265. B
39. C	85. A	131. A	SECTION II	220. D	266. C
40. C	86. D	132. A		221. B	267. C
41. A	87. B	133. C	176. C	222. D	268. A
42. B	88. A	134. D	177. A	223. B	269. B
43. C	89. C	135. A	178. B	224. C	270. D
44. D	90. D	136. B	179. D	225. D	271. C
45. A	91. B	137. C	180. A	226. B	272. B
46. B	92. A	138. A	181. C	227. A	273. C

274. A	326. C	378. D	430. C	482. A	534. C
275. D	327. B	379. A	431. B	483. B	535. B
276. B	328. A	380. B	432. D	484. B	536. D
277. D	329. C	381. D	433. B	485. C	537. A
278. B	330. C	382. A	434. C	486. B	538. B
279. B	331. B	383. A	435. D	487. D	539. C
280. A	332. C	384. C	436. B	488. D	540. B
281. C	333. A	385. D	437. D	489. D	541. A
282. A	334. D	386. C	438. C	490. A	542. D
283. C	335. B	387. B	439. A	491. D	543. B
284. B	336. D	388. B	440. A	492. B	544. C
285. C	337. B	389. C	441. C	493. C	545. D
286. D	338. D	390. D	442. A	494. D	546. A
287. C	339. C	391. A	443. D	495. A	547. B
288. B	340. D	392. D	444. A	496. B	548. A
289. C	341. B	393. B	445. B	497. A	549. A
290. D	342. C	394. B	446. C	498. B	550. D
291. C	343. B	395. A	447. D	499. A	551. D
292. B	344. C	396. C	448. B	500. D	552. C
293. D	345. A	397. C	449. C	501. C	553. B
294. A	346. D	398. B	450. D	502. B	554. B
295. C	347. B	399. D	451. B	503. C	555. B
296. B	348. D	400. A	452. A	504. A	556. B
297. C	349. A	401. B	453. C	505. B	557. A
298. B	350. C	402. D	454. C	506. C	558. B
299. A	351. B	403. A	455. D	507. B	559. D
300. C	352. B	404. C	456. B	508. A	560. A
301. C	353. C	405. A	457. C	509. C	561. C
302. B	354. B	406. D	458. C	510. A	562. D
303. B	355. D	407. B	459. B	511. D	563. C
304. A	356. B	408. A	460. A	512. D	564. A
305. B	357. A	409. A	461. C	513. C	565. C
306. C	358. A	410. B	462. D	514. D	566. A
307. B	359. C	411. D	463. A	515. C	567. D
308. A	360. D	412. D	464. C	516. A	568. A
309. D	361. C	413. D	465. A	517. D	569. D
310. B	362. A	414. B	466. B	518. B	570. B
311. C	363. B	415. C	467. A	519. C	
312. B	364. C	416. A	468. B	520. A	**SECTION III**
313. A	365. D	417. B	469. C	521. B	
314. C	366. C	418. A	470. A	522. D	571. C
315. A	367. D	419. C	471. B	523. C	572. C
316. B	368. B	420. D	472. B	524. A	573. A
317. D	369. B	421. C	473. C	525. C	574. B
318. C	370. C	422. A	474. C	526. B	575. B
319. C	371. A	423. C	475. D	527. C	576. D
320. D	372. B	424. D	476. C	528. A	577. D
321. C	373. C	425. A	477. A	529. C	578. A
322. C	374. B	426. D	478. B	530. A	579. B
323. D	375. A	427. C	479. D	531. B	580. C
324. C	376. D	428. B	480. B	532. D	581. A
325. B	377. C	429. A	481. B	533. A	582. B

583.	C	636.	B	689.	B	742.	A	795.	T	845.	C
584.	C	637.	C	690.	C	743.	C	796.	T	846.	B
585.	C	638.	D	691.	D	744.	A	797.	F	847.	B
586.	D	639.	B	692.	B	745.	D	798.	T	848.	A
587.	B	640.	A	693.	B	746.	C	799.	T	849.	D
588.	A	641.	B	694.	A	747.	B	800.	F	850.	D
589.	C	642.	C	695.	C	748.	A	801.	F	851.	C
590.	D	643.	B	696.	C	749.	A	802.	T	852.	A
591.	B	644.	D	697.	C	750.	C	803.	T	853.	B
592.	B	645.	A	698.	A	751.	A	804.	T	854.	B
593.	A	646.	B	699.	B	752.	B	805.	F	855.	C
594.	B	647.	C	700.	A	753.	A	806.	F	856.	A
595.	A	648.	A	701.	C	754.	B	807.	T	857.	C
596.	C	649.	D	702.	A	755.	D	808.	T	858.	B
597.	B	650.	B	703.	A	756.	C	809.	F	859.	D
598.	A	651.	A	704.	C	757.	D	810.	T	860.	B
599.	B	652.	B	705.	D	758.	A	811.	T	861.	A
600.	C	653.	C	706.	C	759.	B	812.	F	862.	B
601.	B	654.	A	707.	A	760.	C	813.	T	863.	B
602.	B	655.	D	708.	B	761.	B	814.	T	864.	C
603.	C	656.	B	709.	A	762.	C	815.	T	865.	A
604.	C	657.	C	710.	B	763.	A	816.	T	866.	B
605.	A	658.	A	711.	C	764.	D	817.	T	867.	D
606.	A	659.	D	712.	D	765.	B	818.	T	868.	B
607.	C	660.	B	713.	B	766.	A	819.	F	869.	A
608.	B	661.	C	714.	A	767.	B	820.	T	870.	C
609.	A	662.	B	715.	A	768.	A	821.	F	871.	C
610.	B	663.	A	716.	B	769.	B	822.	T	872.	A
611.	D	664.	D	717.	C	770.	D	823.	F	873.	C
612.	B	665.	D	718.	A	771.	C	824.	F	874.	A
613.	C	666.	A	719.	D	772.	B	825.	F	875.	D
614.	D	667.	A	720.	D	773.	C	826.	F	876.	C
615.	A	668.	B	721.	C	774.	A	827.	T	877.	D
616.	B	669.	A	722.	A	775.	B	828.	F	878.	D
617.	C	670.	A	723.	A	776.	A	829.	F	879.	D
618.	A	671.	A	724.	A	777.	D	830.	F	880.	A
619.	C	672.	C	725.	D	778.	A	831.	F	881.	B
620.	D	673.	D	726.	D	779.	A	832.	F	882.	C
621.	B	674.	A	727.	A	780.	B	833.	T	883.	D
622.	A	675.	B	728.	D	781.	D	834.	T	884.	C
623.	C	676.	C	729.	A	782.	A	835.	F	885.	A
624.	A	677.	C	730.	B	783.	B			886.	D
625.	B	678.	C	731.	A	784.	D	SECTION IV		887.	B
626.	D	679.	A	732.	B	785.	B			888.	B
627.	C	680.	D	733.	A	786.	F	836.	D	889.	B
628.	A	681.	B	734.	B	787.	F	837.	A	890.	B
629.	D	682.	C	735.	C	788.	T	838.	D	891.	B
630.	A	683.	C	736.	B	789.	F	839.	C	892.	C
631.	B	684.	A	737.	C	790.	T	840.	A	893.	A
632.	A	685.	C	738.	B	791.	T	841.	B	894.	C
633.	D	686.	B	739.	A	792.	F	842.	C	895.	A
634.	C	687.	C	740.	D	793.	F	843.	B	896.	C
635.	A	688.	B	741.	C	794.	T	844.	D	897.	C

898. C	951. F	1004. F	1054. A	1107. B
899. D	952. T	1005. T	1055. C	1108. C
900. C	953. F		1056. D	1109. B
901. B	954. F	SECTION V	1057. D	1110. B
902. B	955. B		1058. B	1111. C
903. C	956. A	1006. A	1059. B	1112. A
904. B	957. C	1007. A	1060. C	1113. D
905. F	958. D	1008. B	1061. C	1114. C
906. F	959. C	1009. D	1062. A	1115. D
907. D	960. B	1010. A	1063. D	1116. C
908. D	961. A	1011. D	1064. A	1117. B
909. B	962. D	1012. D	1065. A	1118. A
910. C	963. D	1013. A	1066. B	1119. A
911. A	964. C	1014. B	1067. B	1120. D
912. C	965. D	1015. C	1068. D	1121. B
913. C	966. B	1016. A	1069. B	1122. A
914. C	967. B	1017. A	1070. A	1123. C
915. B	968. C	1018. A	1071. D	1124. D
916. D	969. B	1019. B	1072. C	1125. A
917. B	970. A	1020. D	1073. C	1126. D
918. A	971. C	1021. B	1074. A	1127. C
919. C	972. D	1022. D	1075. B	1128. D
920. C	973. C	1023. B	1076. D	1129. A
921. D	974. B	1024. C	1077. C	1130. B
922. C	975. D	1025. A	1078. A	1131. A
923. C	976. B	1026. B	1079. A	1132. A
924. D	977. B	1027. B	1080. D	1133. A
925. D	978. D	1028. A	1081. D	1134. B
926. D	979. B	1029. C	1082. C	1135. D
927. A	980. C	1030. A	1083. D	1136. C
928. C	981. C	1031. B	1084. D	1137. A
929. C	982. A	1032. C	1085. B	1138. B
930. D	983. D	1033. D	1086. C	1139. D
931. C	984. A	1034. A	1087. B	1140. B
932. B	985. D	1035. B	1088. D	1141. D
933. C	986. B	1036. C	1089. C	1142. D
934. B	987. A	1037. B	1090. B	1143. C
935. F	988. C	1038. D	1091. B	1144. B
936. T	989. B	1039. D	1092. D	1145. A
937. T	990. D	1040. C	1093. D	1146. A
938. T	991. D	1041. C	1094. C	1147. A
939. F	992. B	1042. A	1095. B	1148. A
940. T	993. D	1043. D	1096. A	1149. C
941. T	994. D	1044. A	1097. A	1150. B
942. F	995. D	1045. B	1098. D	1151. A
943. T	996. D	1046. D	1099. B	1152. C
944. T	997. T	1047. A	1100. A	1153. D
945. B	998. F	1048. D	1101. C	1154. A
946. C	999. T	1049. B	1102. B	1155. B
947. C	1000. F	1050. D	1103. B	1156. D
948. B	1001. T	1051. B	1104. A	1157. C
949. C	1002. T	1052. C	1105. D	1158. A
950. T	1003. F	1053. A	1106. C	1159. B

ANSWER KEY

1160.	D	**SECTION VI**		**SECTION VII**			
1161.	B						
1162.	C	1213.	A	1263.	D	1316.	A
1163.	C	1214.	C	1264.	B	1317.	B
1164.	A	1215.	B	1265.	A	1318.	A
1165.	D	1216.	A	1266.	B	1319.	B
1166.	A	1217.	D	1267.	A	1320.	B
1167.	D	1218.	B	1268.	B	1321.	A
1168.	B	1219.	D	1269.	A	1322.	B
1169.	A	1220.	A	1270.	A	1323.	A
1170.	C	1221.	A	1271.	B	1324.	B
1171.	D	1222.	C	1272.	C	1325.	B
1172.	A	1223.	B	1273.	D	1326.	B
1173.	D	1224.	D	1274.	A	1327.	C
1174.	C	1225.	C	1275.	A	1328.	C
1175.	A	1226.	C	1276.	C	1329.	Air, water soluble
1176.	B	1227.	B	1277.	D		opaque medium
1177.	B	1228.	D	1278.	B	1330.	Supine
1178.	A	1229.	C	1279.	B	1331.	Auto
1179.	B	1230.	C	1280.	A	1332.	Oblique
1180.	D	1231.	A	1281.	D	1333.	Sialography
1181.	D	1232.	D	1282.	A	1334.	Oily Dionosil
1182.	B	1233.	D	1283.	B	1335.	Selective
1183.	A	1234.	A	1284.	C	1336.	Expiratory
1184.	D	1235.	C	1285.	A	1337.	Left lateral
1185.	D	1236.	D	1286.	C	1338.	F
1186.	B	1237.	C	1287.	A	1339.	T
1187.	T	1238.	D	1288.	B	1340.	T
1188.	F	1239.	B	1289.	A	1341.	F
1189.	T	1240.	D	1290.	B	1342.	T
1190.	F	1241.	A	1291.	A	1343.	T
1191.	T	1242.	A	1292.	B	1344.	F
1192.	F	1243.	B	1293.	C	1345.	T
1193.	F	1244.	C	1294.	A	1346.	F
1194.	T	1245.	C	1295.	D	1347.	T
1195.	F	1246.	B	1296.	A	1348.	T
1196.	T	1247.	B	1297.	D	1349.	T
1197.	F	1248.	D	1298.	B	1350.	F
1198.	T	1249.	B	1299.	D	1351.	T
1199.	F	1250.	D	1300.	B	1352.	F
1200.	F	1251.	B	1301.	A	1353.	B
1201.	T	1252.	A	1302.	C	1354.	A
1202.	F	1253.	D	1303.	D	1355.	K
1203.	T	1254.	B	1304.	A	1356.	D
1204.	F	1255.	C	1305.	C	1357.	I
1205.	T	1256.	A	1306.	A	1358.	G
1206.	T	1257.	C	1307.	B	1359.	H
1207.	F	1258.	D	1308.	D	1360.	C
1208.	T	1259.	C	1309.	D	1361.	J
1209.	F	1260.	C	1310.	A	1362.	B
1210.	F	1261.	A	1311.	A	1363.	D
1211.	F	1262.	D	1312.	B	1364.	D
1212.	F			1313.	C	1365.	B

1366.	C
1367.	B
1368.	A
1369.	B
1370.	B
1371.	B
1372.	A
1373.	C
1374.	D
1375.	B
1376.	B
1377.	D
1378.	A
1379.	A
1380.	C
1381.	B
1382.	B
1383.	C
1384.	B
1385.	D
1386.	D
1387.	A
1388.	A

SECTION VIII

1389.	C
1390.	B
1391.	A
1392.	C
1393.	A
1394.	B
1395.	A
1396.	B
1397.	D
1398.	A
1399.	A
1400.	C
1401.	B
1402.	C
1403.	B
1404.	A
1405.	D
1406.	B
1407.	A
1408.	D
1409.	F
1410.	T
1411.	T
1412.	T
1413.	F
1414.	F
1415.	T

1416.	T
1417.	F
1418.	F
1419.	F
1420.	F
1421.	T
1422.	F
1423.	T
1424.	F
1425.	T
1426.	Dark adaptation
1427.	Visual purple
1428.	Cones
1429.	Opaque
1430.	Illumination
1431.	Inverse Square Law
1432.	Electronic tube
1433.	Resolving power
1434.	X-ray noise
1435.	Periphery brightness falloff
1436.	18 inches
1437.	1 millimeter
1438.	Optical viewer
1439.	Light passing power
1440.	More
1441.	Focussing
1442.	85%
1443.	Frames per second
1444.	40
1445.	16
1446.	Detail
1447.	Speed
1448.	Reversal films
1449.	Maximum

SECTION IX

1450.	D
1451.	A
1452.	A
1453.	B
1454.	C
1455.	C
1456.	B
1457.	D
1458.	B
1459.	A
1460.	B
1461.	D
1462.	D
1463.	A
1464.	A
1465.	C

1466.	B
1467.	B
1468.	B
1469.	D
1470.	A
1471.	C
1472.	C
1473.	B
1474.	B
1475.	C
1476.	A
1477.	B
1478.	A
1479.	C
1480.	B
1481.	C
1482.	A
1483.	C
1484.	B
1485.	B
1486.	B
1487.	B
1488.	D
1489.	C
1490.	A
1491.	D
1492.	C
1493.	A
1494.	B
1495.	D
1496.	C
1497.	B
1498.	B
1499.	D
1500.	D

DATE _____

Please send the following books:

Qty.	Code No.	T I T L E

☐ To save shipping and handling charges my check
is enclosed (including local sales tax).

☐ Bill me. I will remit payment within 30 days,
including mailing charges.

Name _____

Address _____

City _____

State _____ Zip Code _____

(please print)

Other Books Available

NURSING EXAM REVIEW BOOKS

501	Vol. 1	Medical-Surgical Nursing	$ 5.00
502	Vol. 2	Psychiatric-Mental Health Nursing . .	5.00
503	Vol. 3	Maternal-Child Health Nursing	5.00
504	Vol. 4	Basic Sciences	5.00
505	Vol. 5	Anatomy and Physiology	5.00
506	Vol. 6	Pharmacology	5.00
507	Vol. 7	Microbiology	5.00
508	Vol. 8	Nutrition & Diet Therapy	5.00
509	Vol. 9	Community Health	5.00
510	Vol. 10	History and Law of Nursing	5.00
511	Vol. 11	Fundamentals of Nursing	5.00
711	Practical Nursing Exam. Review — Vol. 1		5.00

NURSING OUTLINE SERIES

374	Cancer Nursing	$ 8.00
382	Community Health Nursing	6.00
384	Critical Care Nursing	
392	Gastroenterology Nursing	
388	Gynecologic Nursing	
377	Maternity Nursing	6.00
378	Nursing Fundamentals	6.00
375	Nutrition in Nursing	6.00
381	Orthopedic Nursing	6.00
379	Psychiatric-Mental Health Nursing	6.00

NURSING SELF-ASSESSMENT BOOKS

288	S.A.C.K. in Cardiopulmonary Nursing	$ 8.00
292	S.A.C.K. in Child Health Nursing	8.00
285	S.A.C.K. in Community Health Nursing	8.00
295	S.A.C.K. in Geriatric Nursing	8.00
299	S.A.C.K. in Maternity Nursing	8.00
243	S.A.C.K. in Neurology & Neurosurgical Nursing	8.00
715	S.A.C.K. for the Nurse Anesthetist	8.50

NURSING CASE STUDY BOOKS

036	Maternity Nursing Case Studies	$ 8.50
391	Psych. Comm. Mental Health Nrsg. Case St. .	8.50

MODERN NURSING SERIES

861	Neurology for Nurses	$ 4.00
855	A Handbook for Nurses	4.00
862	The Nursing of Accidents	4.00
866	Emergency and Acute Care	4.00
867	Venereology for Nurses	4.00
868	The Management & Nursing of Burns	4.00
857	Microbiology in Modern Nursing	4.00
856	Obstetrics & Gynecology for Nurses	4.00
858	The Older Patient, An Introduction to Geriatrics	4.00
863	Principles of Medicine & Med. Nursing	4.00
864	Principles of Surgery & Surgery Nursing	4.00
865	Psychology & Psychiatry for Nurses	4.00
859	Communicable Diseases	4.00

OTHER BOOKS FOR NURSES

968	Ambulatory Care Nursing Procedure & Empl. Hlth. Svc. Manual	$12.00
869	Nephrology for Nurses	8.50
376	Nursing and the Nephrology Patient	8.50
358	Renal Transplantation — A Nursing Perspective	8.50

NURSING ESSAY Q & A REVIEW BOOKS

371	Emergency & Disaster Nursing Cont. Ed. Rev. .	$ 8.00
373	Gastroenterology Nursing Continuing Ed. Rev.	8.00
399	Intensive Care Nursing Continuing Ed. Review	6.00
357	Neurology & Neurosurg. Nrsg. Cont. Ed. Rev. .	8.00
356	Nurse Anesthetists Continuing Ed. Review . . .	10.00
350	Obstetric Nursing Continuing Ed. Review	8.00
397	Orthopedic & Rehab. Nursing Cont. Educ. Rev.	8.00
361	Pediatric Nursing Continuing Ed. Review	8.00
351	Psychiatric-Mental Health Nrsg. Cont. Ed. Rev.	8.00
396	Respiratory Nursing Continuing Ed. Review . .	8.00

ALLIED HEALTH REVIEW BOOKS

411	Medical State Board Review — Basic Sciences .	$12.00
412	Medical State Board Review — Clinical Sciences	12.00
473	Cardiopulm. Techn. Exam. Review — Vol. 1 . .	8.50
454	Cytology Exam. Review Book — Vol. 1	8.50
367	Cytology E.R.B. — Vol. 2 (Essay Q & A)	8.50
431	Dental Exam. Review Book — Vol. 1	12.00
432	Dental Exam. Review Book — Vol. 2	12.00
433	Dental Exam. Review Book — Vol. 3	12.00
461	Dental Hygiene Exam. Review — Vol. 1	8.50
465	Emergency Med. Techn. Exam. Rev. — Vol. 1 . .	8.50
466	Emergency Med. Techn. Exam. Rev. — Vol. 2 . .	8.50
424	Immunology Exam. Review Book — Vol. 1	12.00
455	Laboratory Assistants Exam. Rev. Bk. — Vol. 1	8.50
490	Medical Assistants Exam. Rev. Bk. — Vol. 1 . .	8.50
495	Medical Librarian Exam. Rev. Bk. — Vol. 1 . . .	8.50
369	Medical Records Administration Cont. Ed. Rev.	12.00
496	Medical Record Library Science — Vol. 1	8.50
451	Medical Technology Exam. Review — Vol. 1 . .	8.50
452	Medical Technology Exam. Review — Vol. 2 . .	8.50
331	Nuclear Medicine Technology Cont. Ed. Rev. .	12.00
457	Nuclear Medicine Technology Exam. Rev. Book.	12.00
475	Occupational Therapy Exam. Review — Vol. 1 .	8.50
469	Optometry Exam. Review — Vol. 1	12.00
470	Optometry Exam. Review — Vol. 2	12.00
421	Pharmacy Exam. Review Book — Vol. 1	8.50
430	Mill's Pharmacy State Board Q & A	12.00
481	Physical Therapy Exam. Review — Vol. 1	8.50
482	Physical Therapy Exam. Review — Vol. 2	8.50
487	Radiobiology Exam. Review Book	8.50
471	Respiratory Therapy Exam. Review — Vol. 1 . .	8.50
344	Respiratory Therapy Exam. Review — Vol. 2 . .	8.50
423	Sanitarian's Examination Review Book	12.00
441	X-Ray Technology Exam. Review — Vol. 1	8.50
442	X-Ray Technology Exam. Review — Vol. 2	8.50
443	X-Ray Technology Exam. Review — Vol. 3	8.50

TYPIST HANDBOOKS

976	Medical Typist's Guide for Hx & Physicals . . .	$ 6.00
981	Radiology Typist Handbook	6.00
991	Surgical Typist Handbook	6.00
973	Transcribers Guide to Medical Terminology . . .	6.00

LANGUAGE GUIDES

721	English-Spanish Guide for Medical Personnel .	$ 3.00
961	Multilingual Guide for Medical Personnel	4.00
722	Spanish for Hospital Personnel	3.00

Prices subject to change.

P

Other Books Available

Prices subject to change.

MF

Other Books Available

Prices subject to change.   MB